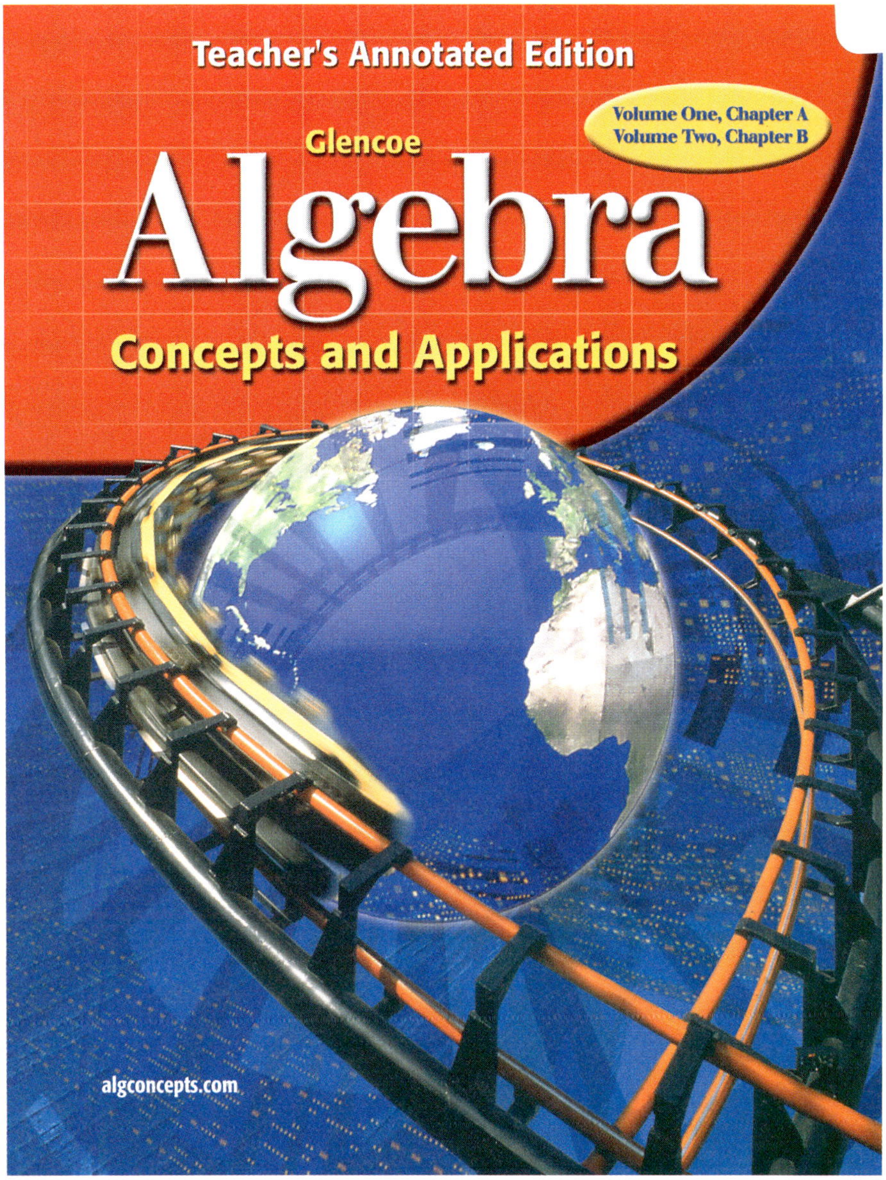

McGraw Hill Glencoe

New York, New York Columbus, Ohio Chicago, Illinois Peoria, Illinois Woodland Hills, California

Glencoe

The McGraw-Hill Companies

Send all inquiries to:
Glencoe/McGraw-Hill
8787 Orion Place
Columbus, OH 43240

Visit our web page at: www.algconcepts.com

ISBN: 0-07-860855-4 (Volume One, Student Edition)
ISBN: 0-07-860775-2 (Volume Two, Student Edition)
ISBN: 0-07-860778-7 (Teacher's Annotated Edition)
Printed in the United States of America.

1 2 3 4 5 6 7 8 9 10 066 12 11 10 09 08 07 06 05 04 03

Glencoe's Algebra 1 in Two Years!

The ever increasing reliance on technology in our world has led to the need for more students to take algebra and geometry in high school. However, some students have difficulty grasping the abstract concepts of algebra. To aid those students, the Algebra 1 curriculum is often taught over a period of two years. **Glencoe's Algebra 1 in Two Years** is designed for such a program. The extended time frame allows students to spend more time on each concept, which aids their development of in-depth understanding. This also gives students more time to complete hands-on labs and activities that develop these concepts.

The **Two-Year Planning Guide and Resources, Year One and Year Two,** are special booklets of teaching aids. They include the Pretests and Posttests from Chapters A and B in the Student Editions, Lesson Planning Guides, Reading to Learn Mathematics masters, and Review Games. These materials are designed to make it easier for you to effectively teach Algebra 1 in two years.

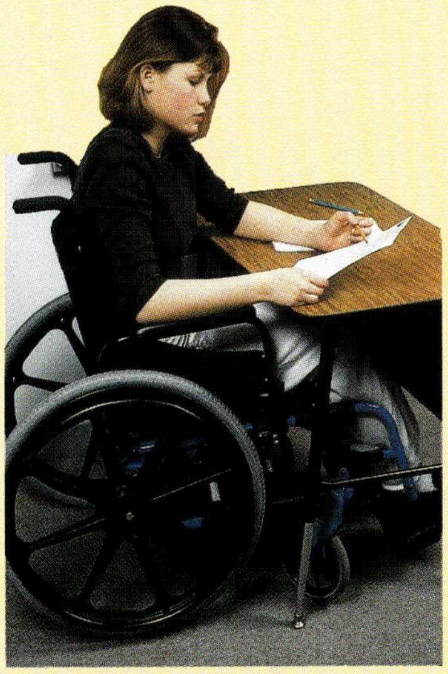

The **Teacher's Classroom Resources** is a comprehensive package of optional materials that support and extend every lesson. This collection includes 7 different blackline masters booklets, a Lesson Planning Guide, Solutions Manual, and other teaching aids. These materials are designed to help you meet the needs of every student.

Table Of Contents

CHAPTER B

Review Lessons by Chapter

COMPONENTS

Glencoe Algebra: Concepts and Applications

Component	ISBN
Volume One, Student Edition	0-07-860855-4
Volume Two, Student Edition	0-07-860775-2
StudentWorks™	0-07-860882-1
Teacher's Wraparound Edition	0-07-845772-6
Teacher's Annotated Edition, Chapters A and B	0-07-860778-7
Two-Year Planning Guide and Resources, Year One	0-07-861193-8
Two-Year Planning Guide and Resources, Year Two	0-07-861194-6
Applications	
School-to-Workplace Masters*	0-07-821543-9
Multimedia Applications CD-ROM	0-07-829064-3
Meeting Individual Needs	
Enrichment Masters*	0-07-821944-2
Practice Masters*	0-07-821544-7
Practice Workbook	0-07-821943-4
Prerequisite Skills Workbook*	0-07-827759-0
Spanish Study Guide and Assessment	0-07-821546-3
Study Guide Masters*	0-07-821547-1
Study Guide Workbook	0-07-821548-X
Technology/Multimedia	
AlgePASS: Concepts and Applications	0-07-821336-3
Graphing Calculator Masters*	0-07-821549-8
Vocabulary PuzzleMaker	0-07-822695-3
Assessment/Evaluation	
Assessment and Evaluation Masters*	0-07-821550-1
MindJogger Videoquizzes	0-07-822696-1
TestCheck and Worksheet Builder	0-07-860885-6
Manipulatives/Modeling	
Overhead Manipulative Resources	0-07-830593-4
Glencoe Mathematics Classroom Manipulative Kit	0-02-833116-8
Glencoe Mathematics Student Manipulative Kit	0-02-833654-2
Hands-On Algebra Masters*	0-07-821551-X
Teaching Aids	
Answer Key Maker	0-07-860883-X
Answer Key Masters	0-07-823093-4
Answer Key Transparencies	0-07-821553-6
Block Schedule Planning Guide*	0-07-821554-4
5-Minute Check Transparencies	0-07-821559-5
Interactive Chalkboard	0-07-860884-8
Lesson Planning Guide*	0-07-821555-2
Solutions Manual*	0-07-821556-0
TeacherWorks™	0-07-860881-3
Teaching Transparencies	0-07-821558-7

Online Study Tools, available at www.algconcepts.com

- Self-Check Quizzes
- Data Updates
- Chapter Investigations
- Math in the Workplace Profiles
- Extra Examples
- Problem-Solving Workshops
- Chapter Review Activities
- Problem of the Week Cards

**Included in the Teacher's Classroom Resources*

PACING CHARTS

Year One		
Chapter	**Days**	
	Regular	**Block**
A*	13	7
1	20	10
2	18	9
3	20	10
4	20	10
5	20	10
6	18	9
7	20	10
8	20	10
Total	**169**	**85**

Year Two		
Chapter	**Days**	
	Regular	**Block**
B*	26	13
8**	20	10
9	16	8
10	16	8
11	20	10
12	20	10
13	20	10
14	16	8
15	18	9
Total	**172**	**86**

*Chapters A and B are optional. See page viii for a description of how to use the pretests and posttests to determine if Chapters A and B are needed.

**Note that if Chapter 8 was covered in Year One, fewer days can be devoted to it in Year Two.

HOW TO USE THE PRETESTS AND POSTTESTS

The pretests and posttests are designed to help teachers determine whether students are ready to proceed with Algebra 1 in Two Years. Teachers should understand that a student's readiness should not be measured solely using one tool. Teachers may also want to consider factors such as classroom performance and the previous school year's math test scores.

Each test is designed to be completed in 35–40 minutes. All of the tests cover the same basic skills necessary for success in Algebra 1 in Two Years.

Calculator use is not assessed in these tests. Therefore, it is not recommended that students use calculators when taking the tests. Computations required in test items are reasonable, and numbers used in computations are restricted.

The table below gives recommendations for evaluating each test and determining a course of action.

Action Plan for Chapter A Pretest and Posttest		
Pretest Score	**Posttest Score**	**Action**
0–14 questions (less than 60%)	0–14 questions (less than 60%)	Present the prerequisite concepts (pages A6–A27).
15–21 questions (60%–84%)	15–21 questions (60%–84%)	Review individual concepts as needed.
22–25 questions (88%–100%)	22–25 questions (88%–100%)	Students are ready to begin the first year of *Algebra 1 in Two Years.*

Action Plan for Chapter B Pretest and Posttest		
Pretest Score	**Posttest Score**	**Action**
0–14 questions (less than 60%)	0–14 questions (less than 60%)	Present the review lessons and/or chapter tests (pages B6–B59).
15–21 questions (60%–84%)	15–21 questions (60%–84%)	Review individual lessons as needed.
22–25 questions (88%–100%)	22–25 questions (88%–100%)	Students are ready to begin the second year of *Algebra 1 in Two Years.*

To The Student

Chapter A contains three sections: pretest, concept review, and posttest. The pretest is a mixture of eleven concepts. These concepts, along with the Prerequisite Skills Review on pages 684–691, are generally acknowledged as necessary for success in *Algebra: Concepts and Applications.* You should take the pretest to determine which concepts you need to work on. The concept review section allows you to develop, and eventually master, the individual skills the pretest identified as needing to be reinforced. You should take the posttest to make sure you understand all of the concepts and to measure your progress.

The chart on the next page contains commonly used formulas and measurement conversions. You may use it on the pretest and posttest.

Formula Chart and Measurement Conversions

Formula Chart		
Perimeter	square rectangle	$P = 4s$ $P = 2(\ell + w)$
Circumference	circle	$C = 2\pi r$
Area	square rectangle triangle trapezoid circle	$A = s^2$ $A = \ell w$ or $A = bh$ $A = \frac{bh}{2}$ or $\frac{1}{2}bh$ $A = \frac{1}{2}h(b_1 + b_2)$ $A = \pi r^2$
Surface Area	cube cylinder (lateral)	$S = 6s^2$ $S = 2\pi rh$
Volume	rectangular prism cylinder cube	$V = \ell wh$ $V = \pi r^2 h$ $V = s^3$
Pythagorean Theorem	right triangle	$a^2 + b^2 = c^2$
Slope of a Line	$m = \frac{y_2 - y_1}{x_2 - x_1}$	
Quadratic Formula	$x = \frac{-b \pm \sqrt{b^2 - 4ac}}{2a}$	
Slope-Intercept Form of an Equation	$y = mx + b$	
Point-Slope Form of an Equation	$y - y_1 = m(x - x_1)$	
Standard Form of an Equation	$Ax + By = C$	

Measurement Conversions		
Measure	**Metric**	**Customary**
Length	1 kilometer = 1000 meters 1 meter = 100 centimeters 1 centimeter = 10 millimeters	1 mile = 1760 yards 1 mile = 5280 feet 1 yard = 3 feet 1 foot = 12 inches
Volume and Capacity	1 liter = 1000 milliliters	1 gallon = 4 quarts 1 gallon = 128 ounces 1 quart = 2 pints 1 pint = 2 cups 1 cup = 8 fluid ounces
Weight and Mass	1 kilogram = 1000 grams 1 gram = 1000 milligrams	1 ton = 2000 pounds 1 pound = 16 ounces
Time		1 year = 12 months 1 year = 52 weeks 1 year = 365 days 1 week = 7 days 1 day = 24 hours 1 hour = 60 minutes 1 minute = 60 seconds

Pretest

Circle the letter of the correct answer.

1. The bar graph shows the approximate times people in the U.S. eat dinner. During which time interval do most people eat dinner?
Source: *USA TODAY* B

Dinner Time

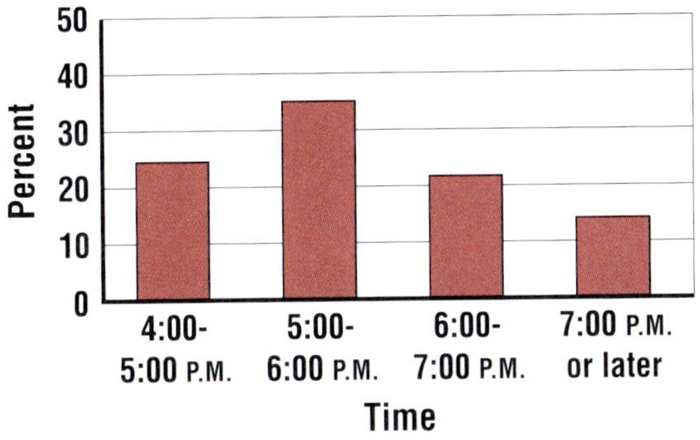

A 4:00–5:00 P.M. **B** 5:00–6:00 P.M.
C 6:00–7:00 P.M. **D** 7:00 P.M. or later

2. On the spinner, there are eight equally likely outcomes. Find the probability of spinning a number greater than 3. H

F $\frac{3}{8}$ **G** $\frac{3}{4}$ **H** $\frac{1}{2}$ **I** $\frac{5}{8}$

3. The line graph shows when people use an automated teller machine. After which day is there the greatest decrease in the number of people using an automated teller machine?
Source: AT&T Global Information Systems D

Automated Teller Machine Usage

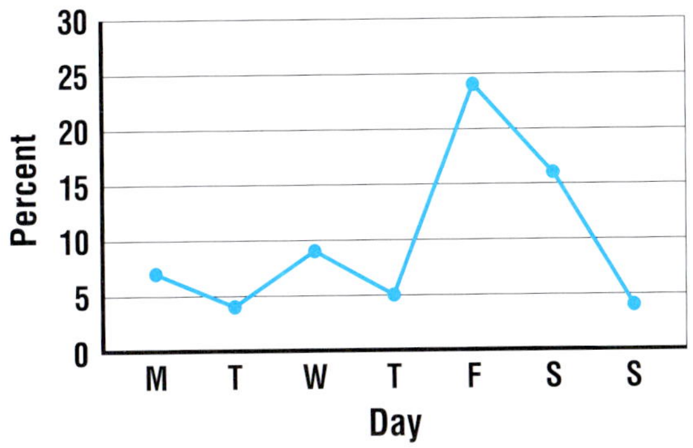

A Tuesday **B** Thursday
C Friday **D** Saturday

4. What is the capacity in liters of a can of soda that contains 355 milliliters of liquid? G

F 3.55 L
G 0.355 L
H 35.5 L
I 35,500 L

5. If a file cabinet drawer is 24 inches deep, how many binders each $1\frac{1}{2}$ inches deep can be used to fill the drawer? C

A 8
B 12
C 16
D 15

6. The circle graph shows how often people barbecue during the summer months. How many more of those surveyed barbecue 2–3 times a month than less than once a month?
Source: American Plastics Council G

Summer Barbecues

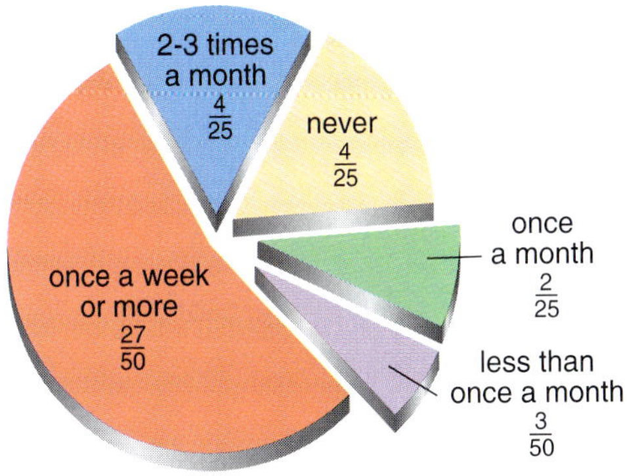

F $\frac{2}{25}$ **G** $\frac{1}{10}$
H $\frac{11}{50}$ **I** $\frac{1}{50}$

7. A full-grown cheetah weighs about 128 pounds. What is the weight of a full-grown cheetah in ounces? D

A 8 oz
B 256 oz
C 1024 oz
D 2048 oz

Pretest

8. Add 3 hours 45 minutes 22 seconds and 5 hours 26 minutes 47 seconds. **G**

F 8 h 22 min 34 s

G 9 h 12 min 9 s

H 9 h 34 min 12 s

I 10 h 1 min 9 s

9. Sean has $8\frac{1}{4}$ yards of fabric to use to make pillows. Each pillow requires $\frac{3}{4}$ yard of fabric. How many pillows can he make? **A**

A 11

B 7

C 9

D 12

10. Find the perimeter of the figure. **I**

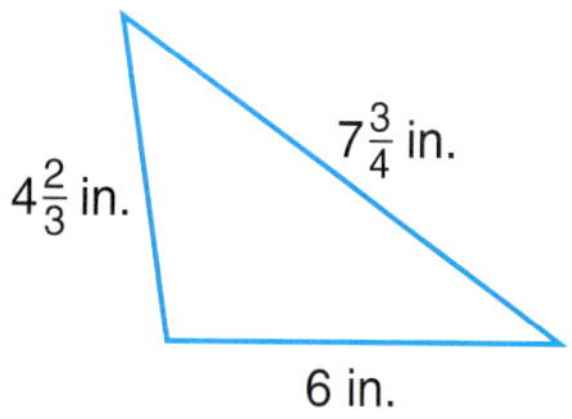

F $17\frac{5}{7}$ in.

G $17\frac{5}{12}$ in.

H $18\frac{3}{4}$ in.

I $18\frac{5}{12}$ in.

11. Find the product of $1\frac{1}{6}$ and $3\frac{3}{4}$. **C**

A $3\frac{1}{8}$

B $3\frac{3}{8}$

C $4\frac{3}{8}$

D $4\frac{7}{8}$

12. What is the area of the figure? **I**

F 49.16 cm^2

G 60.74 cm^2

H 49.64 cm^2

I 60.84 cm^2

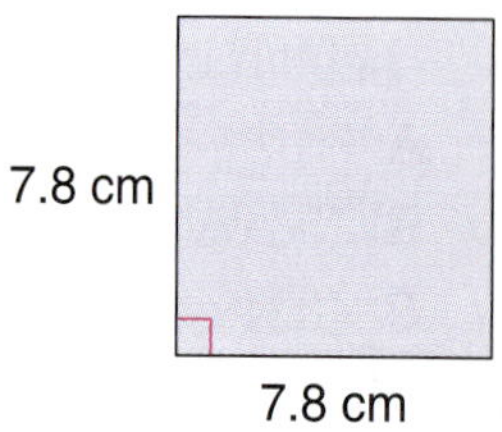

13. Vladimir's dog weighs $11\frac{2}{3}$ pounds. Sara's dog weighs 16 pounds. What is the difference in their weights? **A**

A $4\frac{1}{3}$ lb

B $4\frac{2}{3}$ lb

C $5\frac{1}{3}$ lb

D $5\frac{2}{3}$ lb

14. The time it takes Juanita to walk 7 miles is 123.55 minutes. What is her average time per mile? **G**

F 17.55 min

G 17.65 min

H 17.75 min

I 17.85 min

15. A recipe calls for $\frac{3}{8}$ cup of milk and $\frac{3}{8}$ cup of syrup. What is the total amount of milk and syrup in the recipe? **B**

A $\frac{3}{8}$ c

B $\frac{3}{4}$ c

C $1\frac{1}{8}$ c

D $\frac{9}{64}$ c

16. Jane and Julie are identical twins. At birth, Jane weighed $5\frac{5}{8}$ pounds, and Julie weighed 7 pounds. What was the difference in their weight? **H**

F $\frac{3}{4}$ lb

G $1\frac{1}{4}$ lb

H $1\frac{3}{8}$ lb

I $2\frac{3}{4}$ lb

Pretest

17. In the figure below, what is the total number of rays in which the endpoint is *P*? **D**

A 1
B 2
C 3
D 4

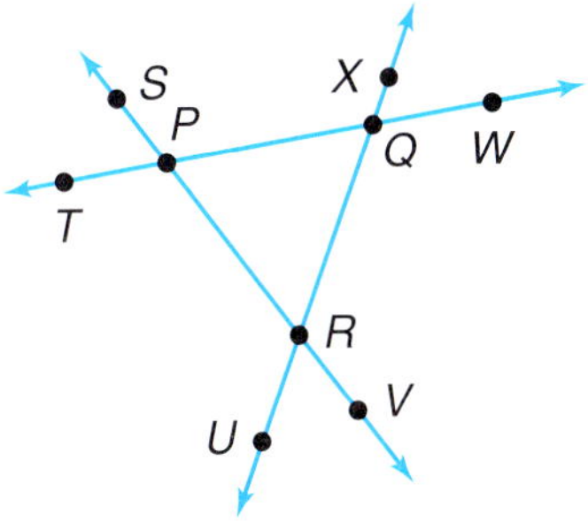

18. The sides of the triangle are best described as: **G**

F rays
G line segments
H angles
I lines

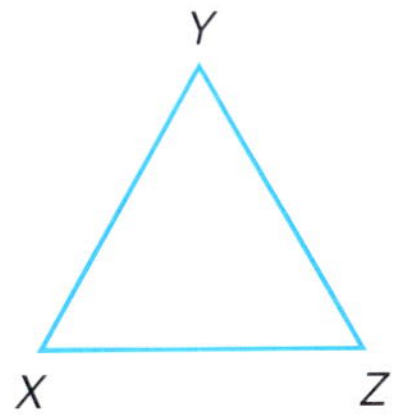

19. Tashema wants to frame a rectangular picture that is 12 inches by 18 inches. What is the perimeter in *feet*? **B**

A 60 ft
B 5 ft
C 24 ft
D $2\frac{1}{2}$ ft

20. What is the surface area of the prism? **H**

F 31.5 cm^2
G 297 cm^2
H 316 cm^2
I 346.5 cm^2

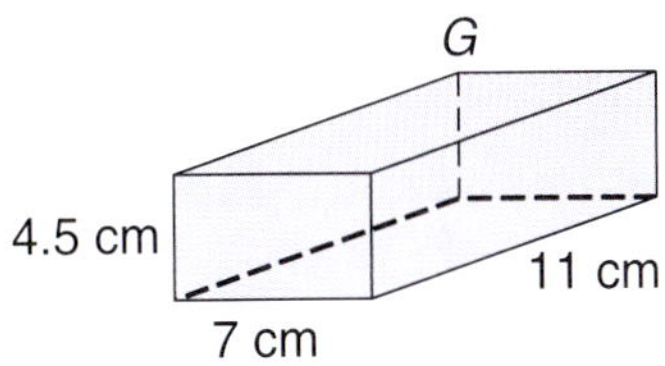

21. A bag contains 3 blue marbles, 6 red marbles, 4 yellow marbles, and 3 black marbles. If you select one marble at random, what is the probability of selecting a yellow marble? **C**

A $\frac{1}{8}$
B $\frac{1}{3}$
C $\frac{1}{4}$
D $\frac{3}{8}$

22. Find the average of Jasmine's test scores. **G**
Jasmine's Test Scores: 85, 93, 79, 88, 91

F 86
G 87.2
H 88
I 89.4

23. What is the volume of the rectangular prism? **D**

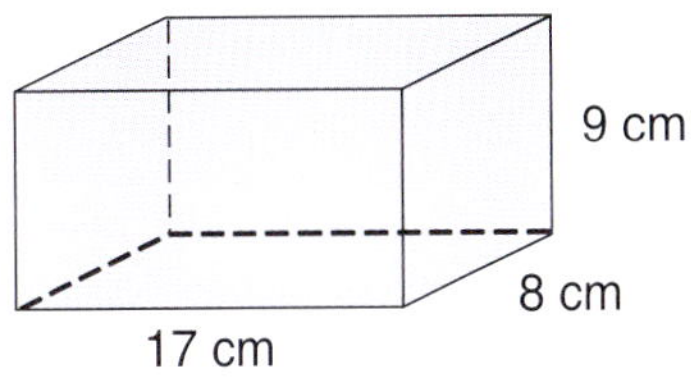

A 204 cm^3
B 722 cm^3
C 1152 cm^3
D 1224 cm^3

24. What is the elapsed time from 9:15 A.M. to 2:40 P.M.? **H**

F 3 hours 15 minutes
G 4 hours 35 minutes
H 5 hours 25 minutes
I 5 hours 35 minutes

25. Tyler ran 2.5 miles after school. How many yards did Tyler run? **A**

A 4400 yards
B 4450 yards
C 5280 yards
D 13,200 yards

PREREQUISITE CONCEPT 1

Reading and Interpreting Graphs

What You'll Learn

You'll learn to read and interpret charts and graphs.

When data are displayed in a chart or graph, you can look for trends and often make decisions or predictions based on the data.

The most common graphs include *line graphs*, *circle graphs*, and *bar graphs*. A **line graph** shows how values change over a period of time.

Examples

1 **In the line graph, what month has the highest normal temperature?**

Locate the highest point of the line graph. Follow the vertical line at that point to the horizontal axis to identify the month.

July has the highest normal temperature.

Raleigh, NC, Monthly Normal Temperature

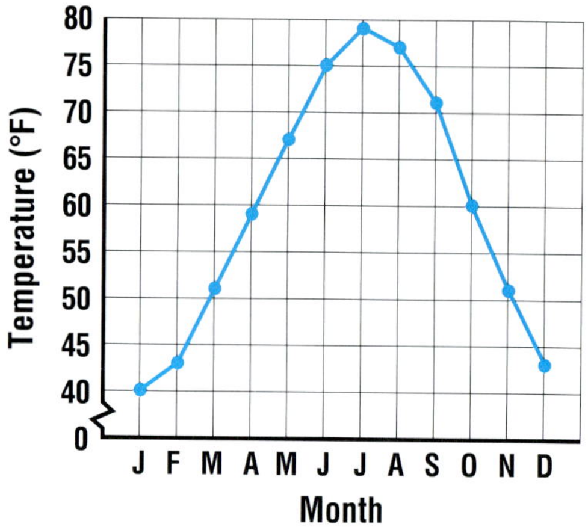

Source: *The World Almanac*

2 **What trend occurs in the normal temperature throughout the year?**

Examine the line graph. The line increases and then decreases. So, the trend in the normal temperature is a gradual increase to higher temperatures in the summer months and then a gradual decrease to lower temperatures as winter approaches.

Circle graphs are used to show how parts are related to the whole.

Example

3 **In the circle graph, what answer did about twice as many people give as *fair/poor*?**

13.7%, or about 14%, said *fair/poor*.

Twice as many people would be 14% × 2, or about 28%.

29.9% of the people surveyed chose *good*, which is close to 28%.

Therefore, about twice as many people surveyed described their health as *good* than those that described their health as *fair/poor*.

How's Your Health?

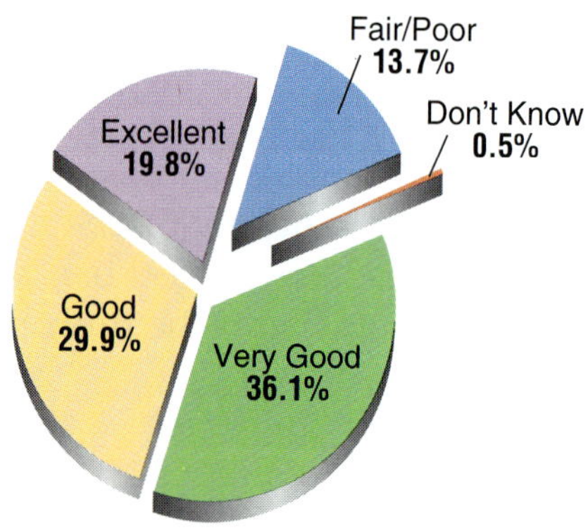

Source: NFO Research, Inc., HealthMed Division

A **bar graph** is used to compare quantities.

Example 4 **Look at the bar graph. For which activity will you burn the most calories in an hour?**

The highest bar on the graph is for jogging. So you will burn the most calories jogging for an hour.

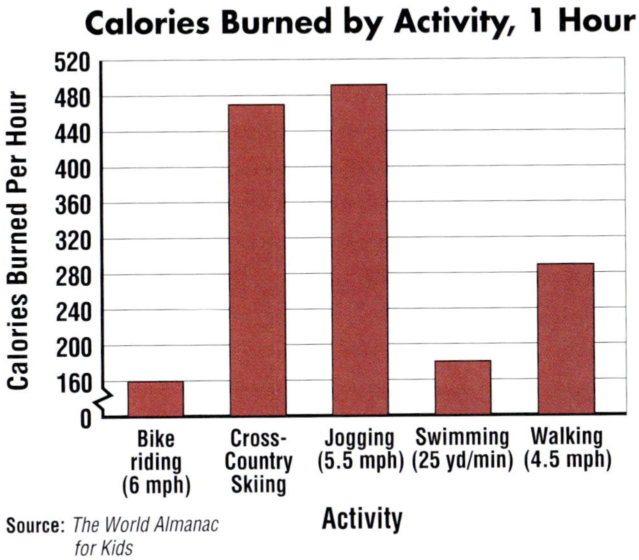

The bar graph shows the number of gold medals the U.S. won at the Winter Olympics from 1984–2002.

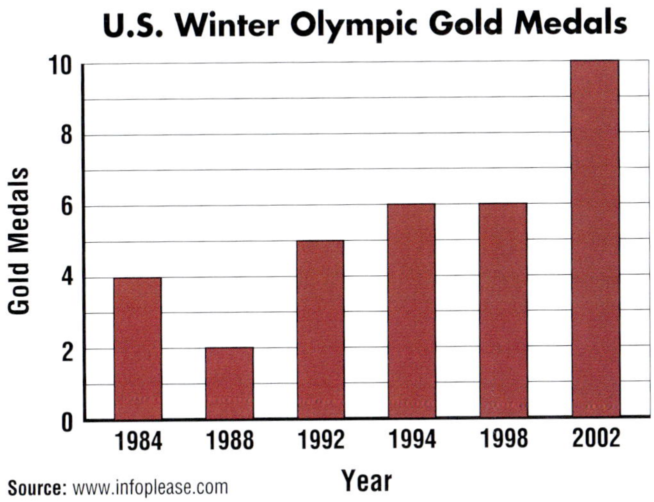

1. In which year did the U.S. win the fewest gold medals? **1998**
2. In which year did the U.S. win the most gold medals? **2002**
3. How many total gold medals did the U.S. win at the Winter Olympics from 1984–2002? **33**

The circle graph shows where the U.S. government spent its money in 2000.

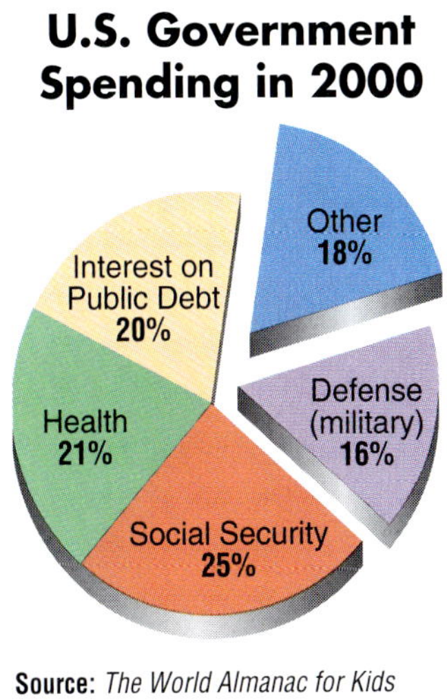

4. Where did the U.S. government spend the most money in 2000? **Social Security**
5. Where did the U.S. government spend $\frac{1}{5}$ of its money in 2000?
6. What fraction of its money did the U.S. government spend on military in 2000? $\frac{4}{25}$

5. interest on public debt

PREREQUISITE CONCEPT 2

Multiplying and Dividing Mixed Numbers

What You'll Learn

You'll learn to multiply and divide mixed numbers.

A **mixed number** is the sum of a whole number and a fraction. To multiply mixed numbers, rename each mixed number as an improper fraction. Then multiply the fractions.

Examples

1 **Find $1\frac{3}{5} \cdot 3\frac{4}{7}$.**

$1\frac{3}{5} \cdot 3\frac{4}{7} = \frac{8}{5} \cdot \frac{25}{7}$ *Express $1\frac{3}{5}$ and $3\frac{4}{7}$ as improper fractions.*

$= \frac{8}{\cancel{5}_1} \cdot \frac{\cancel{25}^5}{7}$ *Divide 5 and 25 by the GCF, 5. Then multiply.*

$= \frac{40}{7}$ or $5\frac{5}{7}$

2 **A cookie recipe calls for $2\frac{1}{4}$ cups of flour. If Dante is tripling the recipe, how many cups of flour will he need?**

Multiply $2\frac{1}{4}$ by 3.

$2\frac{1}{4} \cdot 3 = \frac{9}{4} \cdot \frac{3}{1}$ *Express $2\frac{1}{4}$ as an improper fraction and 3 as $\frac{3}{1}$.*

$= \frac{27}{4}$ or $6\frac{3}{4}$

Dante will need $6\frac{3}{4}$ cups of flour.

To divide by a mixed number, rename the mixed number as an improper fraction. Then multiply by its multiplicative inverse.

Example

3 **Find $\frac{2}{3} \div 2\frac{6}{7}$.**

$\frac{2}{3} \div 2\frac{6}{7} = \frac{2}{3} \div \frac{20}{7}$ *Express $2\frac{6}{7}$ as an improper fraction.*

$= \frac{2}{3} \cdot \frac{7}{20}$ *Dividing by $\frac{20}{7}$ is the same as multiplying by $\frac{7}{20}$.*

$= \frac{\cancel{2}^1}{3} \cdot \frac{7}{\cancel{20}_{10}}$ *Divide 2 and 20 by the GCF, 2. Then multiply.*

$= \frac{7}{30}$

Example 4 **Tory is planning a pizza party for herself and 8 of her friends. She estimates that each person will eat $\frac{2}{5}$ of a pizza. If she buys 4 pizzas, will there be enough pizza?**

Divide the number of pizzas by the estimated amount that each person will eat.

$$4 \div \frac{2}{5} = \frac{4}{1} \cdot \frac{5}{2}$$
$$= \frac{20}{2}$$
$$= 10$$

Based on her estimate, Tory will have enough pizza to feed 10 people. So she will have enough pizza for herself and her 8 friends.

Multiply. Write each product in simplest form.

1. $2 \cdot 1\frac{5}{18}$ $2\frac{5}{9}$
2. $3\frac{2}{5} \cdot 10$ 34
3. $2 \cdot 3\frac{5}{8}$ $7\frac{1}{4}$
4. $2\frac{2}{3} \cdot \frac{4}{5}$ $2\frac{2}{15}$
5. $\frac{7}{8} \cdot 4\frac{1}{4}$ $3\frac{23}{32}$
6. $2\frac{3}{10} \cdot \frac{5}{12}$ $\frac{23}{24}$
7. $9\frac{1}{3} \cdot 3\frac{3}{4}$ 35
8. $3\frac{1}{3} \cdot 4\frac{1}{2}$ 15
9. $6\frac{1}{4} \cdot 1\frac{7}{16}$ $8\frac{63}{64}$
10. $1\frac{7}{8} \cdot 4\frac{2}{7}$ $8\frac{1}{28}$
11. $3\frac{1}{8} \cdot 2\frac{4}{5}$ $8\frac{3}{4}$
12. $3\frac{2}{3} \cdot 1\frac{1}{11}$ 4

Divide. Write each quotient in simplest form.

13. $5 \div 1\frac{1}{3}$ $3\frac{3}{4}$
14. $\frac{3}{4} \div 1\frac{1}{2}$ $\frac{1}{2}$
15. $2\frac{1}{4} \div \frac{2}{3}$ $3\frac{3}{8}$
16. $13 \div 2\frac{3}{5}$ 5
17. $2\frac{1}{4} \div 1\frac{2}{7}$ $1\frac{3}{4}$
18. $5\frac{1}{4} \div 3\frac{1}{2}$ $1\frac{1}{2}$
19. $1\frac{3}{4} \div \frac{7}{12}$ 3
20. $\frac{3}{5} \div 11\frac{2}{5}$ $\frac{1}{19}$
21. $4\frac{2}{5} \div \frac{11}{15}$ 6
22. $1 \div 2\frac{3}{5}$ $\frac{5}{13}$
23. $2\frac{5}{8} \div 7\frac{1}{2}$ $\frac{7}{20}$
24. $5\frac{3}{5} \div 4\frac{1}{5}$ $1\frac{1}{3}$
25. **Economics** The sales tax in Franklin County is $7\frac{3}{4}$%. This means for every dollar you spend, you pay $7\frac{3}{4}$¢ in tax. If you purchase an item for $24, how much will you owe in tax? **$1.86**
26. **Carpentry** Builders Square has two stacks of $1\frac{1}{4}$-inch plywood sheets. One stack is 80 inches high. The other stack is 45 inches high. How many sheets of plywood are there in both stacks? **100**

PREREQUISITE CONCEPT 3

Adding and Subtracting Mixed Numbers

What You'll Learn

You'll learn to add and subtract mixed numbers.

To add or subtract mixed numbers:

1. Add or subtract the fractions. If necessary, rename the fractions first.
2. Add or subtract the whole numbers.
3. Simplify.

Examples

Find each sum.

1 $2\frac{3}{4} + 5\frac{1}{8}$

$2\frac{3}{4} + 5\frac{1}{8} = 2\frac{6}{8} + 5\frac{1}{8}$	*Rename $\frac{3}{4}$ as $\frac{6}{8}$.*
$= (2 + 5) + \left(\frac{6}{8} + \frac{1}{8}\right)$	*Add the whole numbers and the fractions.*
$= 7 + \frac{7}{8}$	*Simplify.*
$= 7\frac{7}{8}$	*Add.*

2 $6\frac{3}{5} + 9\frac{1}{2}$

$6\frac{3}{5} + 9\frac{1}{2} = 6\frac{6}{10} + 9\frac{5}{10}$	*Use the LCM of 5 and 2 to rename $\frac{3}{5}$ as $\frac{6}{10}$ and $\frac{1}{2}$ as $\frac{5}{10}$.*
$= 15\frac{11}{10}$	*The mixed number contains an improper fraction.*
$= 15 + \frac{11}{10}$	*Rewrite as the sum of the whole number and improper fraction.*
$= 15 + 1\frac{1}{10}$	*Rename $\frac{11}{10}$ as a mixed number.*
$= 16\frac{1}{10}$	*Add.*

Example

3 **Find $8\frac{2}{3} - 6\frac{3}{7}$.**

$8\frac{2}{3} - 6\frac{3}{7} = 8\frac{14}{21} - 6\frac{9}{21}$	*Use the LCM of 3 and 7 to rename $\frac{2}{3}$ and $\frac{3}{7}$.*
$= (8 - 6) + \left(\frac{14}{21} - \frac{9}{21}\right)$	*Subtract the whole numbers and the fractions.*
$= 2 + \frac{5}{21}$	*Simplify.*
$= 2\frac{5}{21}$	*Add.*

Sometimes, when you subtract two mixed numbers, the fraction in the first mixed number is less than the fraction in the second. In this case, you need to rename the first mixed number before subtracting.

Example 4 Find $9\frac{1}{3} - 7\frac{5}{6}$.

$9\frac{1}{3} - 7\frac{5}{6} = 9\frac{2}{6} - 7\frac{5}{6}$

$= \left(8 + 1 + \frac{2}{6}\right) - 7\frac{5}{6}$ *Rename $9\frac{2}{6}$ so the fraction is bigger than $\frac{5}{6}$.*

$= \left(8 + \frac{6}{6} + \frac{2}{6}\right) - 7\frac{5}{6}$ *Rename 1 as $\frac{6}{6}$.*

$= 8\frac{8}{6} - 7\frac{5}{6}$ *Add.*

$= 1\frac{3}{6}$ or $1\frac{1}{2}$ *Subtract.*

Add or subtract. Write the solution in simplest form.

1. $6\frac{3}{4} - 6\frac{1}{4}$ $\frac{1}{2}$
2. $10\frac{5}{6} + 5\frac{1}{6}$ 16
3. $4\frac{3}{4} - 1\frac{3}{8}$ $3\frac{3}{8}$
4. $4\frac{1}{2} + \frac{5}{8}$ $5\frac{1}{8}$
5. $7\frac{5}{8} - 4\frac{5}{8}$ 3
6. $4\frac{5}{8} - 3\frac{1}{2}$ $1\frac{1}{8}$
7. $73\frac{1}{2} + \frac{2}{3}$ $74\frac{1}{6}$
8. $3\frac{1}{4} + 5\frac{1}{3}$ $8\frac{7}{12}$
9. $15 - 2\frac{1}{3}$ $12\frac{2}{3}$
10. $8\frac{7}{8} - 4\frac{2}{3}$ $4\frac{5}{24}$
11. $9\frac{1}{3} + 5\frac{5}{6}$ $15\frac{1}{6}$
12. $4\frac{1}{2} - 2\frac{3}{4}$ $1\frac{3}{4}$
13. $4\frac{1}{2} + 7\frac{5}{8}$ $12\frac{1}{8}$
14. $5\frac{5}{16} - 2\frac{3}{8}$ $2\frac{15}{16}$
15. $1\frac{2}{3} + 5\frac{7}{8}$ $7\frac{13}{24}$

16. **Sewing** Becky is making a suit. She needs $2\frac{5}{8}$ yards of material for the jacket and $\frac{7}{8}$ yard of material for the skirt. What is the minimum amount of material Becky needs to make the suit? $3\frac{1}{2}$ yards

17. **Stock Market** On Monday, the stock of a telephone company was at $21\frac{3}{8}$. By Friday, the stock had increased to $35\frac{5}{16}$. By how much did the stock increase over the week? $13\frac{15}{16}$

PREREQUISITE CONCEPT 4

Finding Simple Probability

What You'll Learn

You'll learn to find the probability of a simple event.

An **outcome** is a possible result of something occurring. For example, 6 is an outcome of rolling a number cube. When all outcomes are equally-likely to occur, the outcomes happen at **random**.

An **event** is a specific outcome or type of outcome. For example, selecting a red jelly bean from a bag at random is an event.

The **probability** of an event is the chance that an event will happen.

Probability	
Words:	The probability of an event is the ratio of the number of ways an event can occur to the number of possible outcomes.
Symbols:	$P(\text{event}) = \dfrac{\text{number of ways event occurs}}{\text{number of possible outcomes}}$

The probability of an outcome is a number from 0 to 1.

- A probability of 0 means that an event is impossible.
- A probability of 1 means that an event is certain to happen.
- The closer a probability is to 1, the more likely the event is to happen.

Example 1

A spinner like the one at the right is used in a game. What is the probability that the spinner will stop on blue when you spin it?

There is one way to stop on blue.

There are 4 possible outcomes.

$$P(\text{blue}) = \frac{\text{number of ways to stop on blue}}{\text{number of possible outcomes}} = \frac{1}{4}$$

Examples

A bag contains five yellow marbles, four red marbles, two blue marbles, and one green marble.

2 What is the probability of selecting a red marble?

$$P(\text{red}) = \frac{\text{number of red marbles}}{\text{total number of marbles}} = \frac{4}{5+4+2+1} = \frac{4}{12} \text{ or } \frac{1}{3}$$

3 **What is the probability of selecting a blue or green marble?**

$$P(\text{blue or green}) = \frac{\text{number of blue and green marbles}}{\text{total number of marbles}}$$

$$= \frac{2 + 1}{12} \text{ or } \frac{1}{4}$$

Find the number of possible outcomes for each event.

1. a coin is tossed 2
2. a number cube is rolled 6
3. a spinner with 5 equal sections is spun 5
4. two number cubes are rolled and the outcome is the sum of the number rolled 11

A drawer contains one red sock, three blue socks, two black socks, and two green socks. One sock is removed. Find each probability.

5. $P(\text{red})$ $\frac{1}{8}$
6. $P(\text{blue})$ $\frac{3}{8}$
7. $P(\text{not black})$ $\frac{3}{4}$
8. $P(\text{green})$ $\frac{1}{4}$
9. $P(\text{orange})$ 0
10. $P(\text{red or blue})$ $\frac{1}{2}$

The spinner at the right is spun once. Find each probability.

11. $P(3)$ $\frac{1}{12}$
12. $P(5 \text{ or } 10)$ $\frac{1}{6}$
13. $P(\text{even})$ $\frac{1}{2}$
14. $P(\text{odd})$ $\frac{1}{2}$
15. $P(\text{less than one})$ 0
16. $P(\text{prime})$ $\frac{5}{12}$
17. $P(\text{multiple of 3})$ $\frac{1}{3}$
18. $P(\text{greater than 9})$ $\frac{1}{4}$

19. **Retail Sales** Mercedes surveyed 75 students to determine how many of each size of school jackets to order. The results of the survey are shown in the table at the right.

19a. $\frac{1}{5}$

a. If one of the students is chosen at random, what is the probability that the student wears a medium?

b. If one of the students is chosen at random, what is the probability that the student wears a large or extra-large? $\frac{8}{15}$

Size	Tally
Small	𝍸 𝍸 𝍸 𝍸
Medium	𝍸 𝍸 𝍸
Large	𝍸 𝍸 𝍸 𝍸 III
Extra-large	𝍸 𝍸 𝍸 II

20. $\frac{1}{58}$

20. **Fund-Raiser** For a fund-raiser, the school band is selling raffle tickets for a DVD player. They sold a total of 870 tickets. If your family bought 15 tickets, what is the probability that your family will win the DVD player?

21. In Exercise 20, what is the probability that your family will not win the DVD player? $\frac{57}{58}$

PREREQUISITE CONCEPT 5

Finding Averages

What You'll Learn

You'll learn to find the average of a set of data.

The **average** of a set of data is a middle measure that is found by adding the numbers in the data set and dividing by the number of items in the set. For example, your grade for a class might be the average of your test scores throughout the quarter. Therefore it can be found by adding your test scores and dividing by the total number of tests.

Example

1 **Micah spent $17.10 on lunch this week in the cafeteria. If he bought his lunch 5 days this week, what is the average amount he spent on lunch each day?**

Since you already know the total amount Micah spent on lunch for the week, divide by the number of days he bought his lunch.

$$\frac{\$17.10}{5} = \$3.42$$

Micah spent an average of $3.42 each day for lunch.

Examples

Find the average of each set of data.

2 **35, 18, 26, 15, 42**

$$\begin{aligned} \text{average} &= \frac{\text{sum of values}}{\text{number of values}} \\ &= \frac{35 + 18 + 26 + 15 + 42}{5} \\ &= \frac{136}{5} \text{ or } 27.2 \end{aligned}$$

The average of the set of data is 27.2.

3 **17, 14, 12, 15, 16, 14, 11, 17, 16, 18**

$$\begin{aligned} \text{average} &= \frac{17 + 14 + 12 + 15 + 16 + 14 + 11 + 17 + 16 + 18}{10} \\ &= \frac{150}{10} \text{ or } 15 \end{aligned}$$

The average of the set of data is 15.

4 **3.5, 4.2, 7.6, 5.1, 3.8, 4.1, 3.9**

$$\begin{aligned} \text{average} &= \frac{3.5 + 4.2 + 7.6 + 5.1 + 3.8 + 4.1 + 3.9}{7} \\ &= \frac{32.3}{6} \text{ or } 4.6 \end{aligned}$$

The average of the set of data is 4.6.

1. **Travel** Michael drove 513 miles in 9 hours. What was Michael's average speed? 57 mph

2. **Health** Mirassa ate a total of 1458 Calories in three meals today. What was the average number of Calories she ate in each meal? 486 Calories

Find the average of each set of data. If necessary, round to the nearest hundredth.

3. 5, 3, 8, 3, 6, 4, 6 5

4. 48, 40, 43, 38, 27, 51, 47 42

5. 90, 85, 83, 92, 76 85.2

6. 1.8, 2.2, 2.3, 1.7, 0.6, 0.2 1.47

7. 29, 54, 16, 54, 84, 53, 41, 35, 46, 84, 53, 15 47

8. 1.25, 1.43, 1.5, 1.74, 1.86, 1.05, 1.34, 1.11 1.41

9.

Student	Height (in.)
Abby	62
Carlton	67
Darnell	69
Kim	63
Niko	70
Simon	66
Yolanda	58

65 in.

10.

Day	Hours Worked
Sunday	5.5
Tuesday	8
Wednesday	6.5
Thursday	3.5
Saturday	7

6.1 h

11.

Test	Grade (%)
Chapter 1	78
Chapter 2	83
Chapter 3	88
Chapter 4	80
Chapter 5	91

84%

12.

Day of Vacation	Amount Spent on Meals ($)
Sunday	48.55
Monday	26.70
Tuesday	98.32
Wednesday	50.16
Thursday	25.18
Friday	73.27
Saturday	34.68

$50.98

13. 1162.5 mi

Month	January	February	March	April	May	June
Miles Driven	1052	1308	1205	945	1053	1106
Month	July	August	September	October	November	December
Miles Driven	1874	1351	874	1094	1132	956

14. **Sports** The number of points Drew scored in each basketball game this season is 5, 11, 8, 16, 3, 6, 8, 5, 2, 12, 10, 7, 13, 15, and 11. What was Drew's average number of points per game for the season? 8.8 points

PREREQUISITE CONCEPT 6 Metric Measurement

What You'll Learn

You'll learn to use and convert metric measurements.

The **metric system** is a base-ten system of weights and measures.

The **meter** is the basic unit of length. All lengths in the metric system are described in terms of the meter.

The **kilogram** is the basic unit of mass. All mass in the metric system is described in terms of the gram.

The **liter** is the basic unit of capacity. All capacity in the metric system is described in terms of the liter.

To change from one unit to another within the metric system, you either multiply or divide by powers of 10. The chart below shows the relationship between the units in the metric system and the powers of 10.

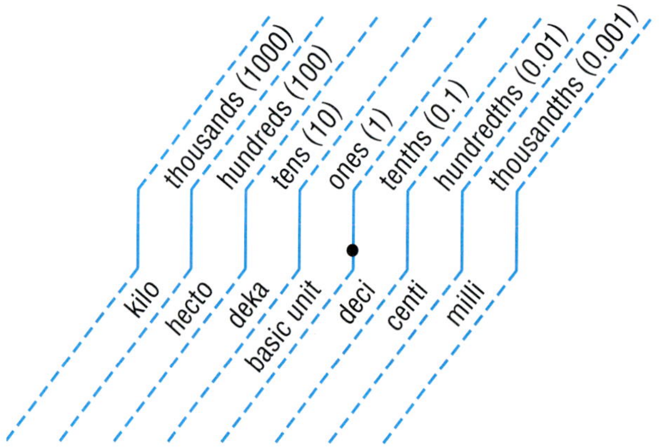

Notice that each place value is 10 times the place value to the right.

To change from a larger unit to a smaller unit, you need to multiply. To change from a smaller unit to a larger unit, you need to divide.

The diagram can help you change among the more common metric units.

MULTIPLY to change from larger units to smaller units.

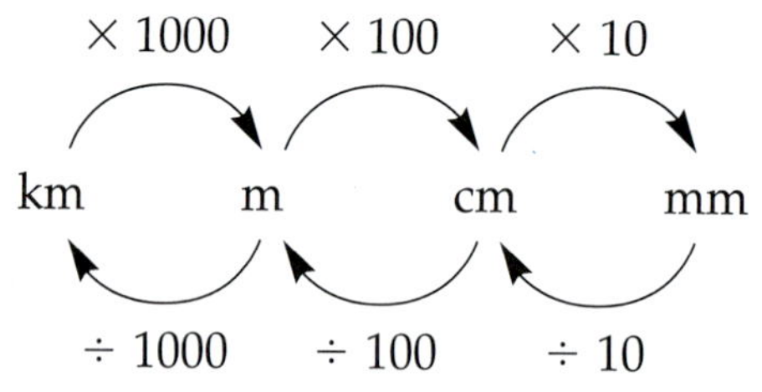

DIVIDE to change from smaller units to larger units.

Examples

Complete each sentence.

1 **200 cm = __?__ m**

Meters are larger than centimeters, so you are converting to larger units. Divide by 100 since 1 m = 100 cm.

$200 \div 100 = 2$

200 cm = 2 m

2 **0.354 L = ? mL**

Milliliters are smaller than liters, so you are converting to smaller units. Multiply by 1000 since 1 L = 1000 mL.

$0.354 \times 1000 = 354$

0.354 L = 354 mL

3 **13,454 g = ? kg**

Kilograms are larger than grams, so you are converting to larger units. Divide by 1000 since 1 kg = 1000 g.

$13{,}454 \div 1000 = 13.454$

13,454 g = 13.454 kg

Complete each sentence.

1. 0.6 cm = ? mm **6**
2. 300 mL = ? L **0.3**
3. 1495 mg = ? g **1.495**
4. 3.75 kg = ? g **3750**
5. 6.9 mm = ? cm **0.69**
6. 2.5 km = ? m **2500**
7. 9.1 kg = ? mg **9,100,000**
8. 34 cm = ? m **0.34**
9. 642 mL = ? L **0.642**
10. 7 g = ? mg **7000**
11. 906 mL = ? L **0.906**
12. 0.3 cm = ? mm **3**
13. 0.31 g = ? mg **310**
14. 3861 m = ? km **3.861**
15. 9.2 L = ? mL **9200**
16. 1.7035 kL = ? L **1703.5**
17. 27 g = ? mg **27,000**
18. 54 L = ? kL **0.054**
19. 7.913 km = ? cm **791,300**
20. 24,520 mg = ? kg **0.02452**

Which metric unit would you use to measure each item?

21. water in a swimming pool **kL**
22. width of a door **cm, m**
23. weight of a 15-year old male **kg**
24. weight of a soup can **mg**

25. **Food** A small container of yogurt has a mass of 113 grams. The containers are sold in packs of six. What is the mass in kilograms of one six-pack of yogurt? **0.678 kg**

26. **Archaeology** The dimensions of an arrowhead are shown in the illustration at the right. Will this arrowhead fit in a box that is 0.04 m long and 2.5 cm wide? Explain your reasoning. **Yes; the dimensions of the box are 40 mm by 25 mm.**

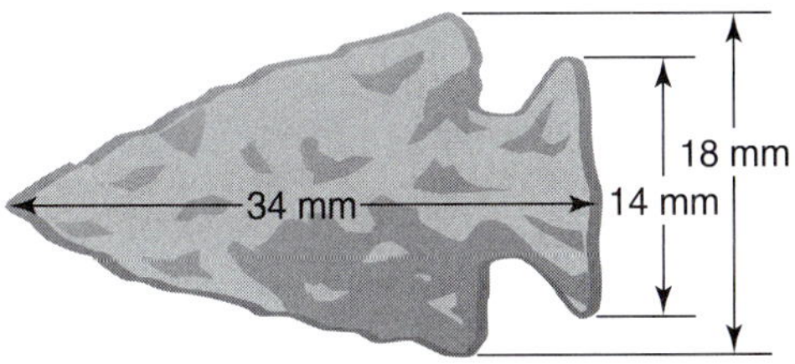

PREREQUISITE CONCEPT 7

Customary Measurement

What You'll Learn

You'll learn to use and convert customary measurements.

The **customary system** is the measurement system that is more commonly used in the United States.

The most commonly used customary units of length are the **inch**, **foot**, **yard**, and **mile**.

Customary Units of Length	
	1 foot (ft) = 12 inches (in.)
	1 yard (yd) = 3 feet or 36 inches
	1 mile (mi) = 5280 feet or 1760 yards

The most commonly used customary units of capacity are the **fluid ounce**, **cup**, **pint**, **quart**, and **gallon**.

Customary Units of Capacity	
	1 cup (c) = 8 fluid ounces (fl oz)
	1 pint (pt) = 2 cups
	1 quart (qt) = 2 pints
	1 gallon (gal) = 4 quarts

The most commonly used customary units of weight are **ounce**, **pound**, and **ton**.

Customary Units of Weight	
	1 pound (lb) = 16 ounces (oz)
	1 ton (T) = 2000 pounds

Converting from one unit to another in the customary system may involve more than one step. The rules about when to multiply or divide are the same as those used in the metric system. To change from a larger unit to a smaller unit, you need to multiply. To change from a smaller unit to a larger unit, you need to divide.

Examples

Complete each sentence.

1 **1035 ft = _?_ yd**

Yards are larger than feet, so you are converting to larger units. Divide by 3 since 1 yd = 3 ft.

$1035 \div 3 = 345$

1035 ft = 345 yd

2 **40 oz = _?_ lb**

Pounds are larger than ounces, so you are converting to larger units. Divide by 16 since 1 lb = 16 oz.

40 ÷ 16 = 2.5

40 oz = 2.5 lb

3 **9 c = _?_ qt**

Quarts are larger than cups, so you are converting to larger units.

First convert cups to pints. Divide by 2 since 1 pt = 2 c.

9 ÷ 2 = 4.5

9 c = 4.5 pt

Then convert pints to quarts. Divide by 2 since 1 qt = 2 pt.

4.5 ÷ 2 = 2.25

9 c = 2.25 qt

Complete each sentence.

1. 12 ft = _?_ in. 144
2. 4 T = _?_ lb 8000
3. 9 pt = _?_ c 18
4. 31,680 ft = _?_ mi 6
5. 3 lb = _?_ oz 48
6. 15 qt = _?_ gal 3.75
7. 6 ft = _?_ in. 72
8. 3 gal = _?_ pt 24
9. 48 ft = _?_ yd 16
10. 2 mi = _?_ yd 3520
11. 7 lb = _?_ oz 112
12. 7000 lb = _?_ T 3.5
13. 6 yd = _?_ in. 216
14. 48 fl oz = _?_ pt 3
15. 12 oz = _?_ lb 0.75
16. 1.4 mi = _?_ ft 7392
17. 20 pt = _?_ gal 2.5
18. 31,680 in. = _?_ mi 0.5

Which metric unit would you use to measure each item?

19. the weight of a baby lb
20. height of a 16-year old female ft
21. water in a fish tank gal
22. distance from home to school mi

23. **Biology** A dinosaur called a brachiosaurus weighed about 170,000 pounds. A blue whale weighs about 220 tons. How many more tons does today's blue whale weigh than the brachiosaurus? 135 tons

PREREQUISITE CONCEPT 8

Units of Time

What You'll Learn

You'll learn to add and subtract units of time.

The most common units of time are **seconds**, **minutes**, and **hours**.

Units of Time	1 hour (h) = 60 minutes (min) 1 minute = 60 seconds (s)

To add or subtract measures of time:

1. Add or subtract the seconds.
2. Add or subtract the minutes.
3. Add or subtract the hours.

Rename if necessary in each step.

Examples

1 **Find each sum.**

6 min 13 s + 11 min 26 s

```
   6 min 13 s      Add vertically.
+ 11 min 26 s      Add the seconds and then the minutes.
  17 min 39 s
```

2 **3 h 42 min + 5 h 35 min**

```
  3 h 42 min      Add vertically.
+ 5 h 35 min      Add the minutes and then the hours.
  8 h 77 min
```

Since the number of minutes is greater than 60, rename 77 min as 1 h and 17 min.

8 h + 1 h 17 min = 9 h 17 min

Example

3 **Find 17 h 42 min − 12 h 50 min.**

```
  17 h 42 min      Subtract vertically.
− 12 h 50 min
```

Since you cannot subtract 50 min from 42 min, you must rename 17 h 42 min as 16 h (42 + 60) min, or 16 h 102 min.

```
  16 h 102 min
− 12 h  50 min
   4 h  52 min
```

Example 4 **Travel** **Kayla's flight is scheduled to leave Atlanta at 11:40 A.M. and arrive in Dallas at 2:04 P.M. How long will the flight last?**

You need to find the elapsed time from 11:40 A.M. to 2:04 P.M.

From 11:40 A.M. to 12:00 noon is 20 minutes.
From 12:00 noon to 2:04 P.M. is 2 hours 4 minutes.

The elapsed time is 20 min + 2 h 4 min, or 2 h 24 min.
Kayla's flight will last 2 hours and 24 minutes.

Add or subtract.

1. 7 min 13 s − 4 min 10 s = **3 min 3 s**	2. 10 h 33 min + 9 h 12 min = **19 h 45 min**	3. 2 min 13 s + 4 min 6 s = **6 min 19 s**
4. 9 h 59 s − 7 h 37 s = **2 h 22 s**	5. 48 min 22 s − 46 min 35 s = **1 min 47 s**	6. 1 h 54 min + 47 min = **2 h 41 min**
7. 13 h 6 min − 10 h 41 min = **2 h 25 min**	8. 9 h 37 s − 6 h 9 s = **3 h 28 s**	9. 5 h 29 min + 3 h 31 min = **9 h**
10. 6 h 37 min 22 s + 5 h 22 min 16 s = **11 h 59 min 38 s**	11. 5 h 30 s − 3 h 8 min 25 s = **1 h 52 min 5 s**	12. 6 h + 10 h 13 min = **16 h 13 min**
13. 3 h − 1 h 18 min 25 s = **1 h 41 min 35 s**	14. 11 h 52 min 17 s + 6 h 13 min 50 s = **18 h 6 min 7 s**	15. 2 h 12 min 33 s − 1 h 24 min 47 s = **47 min 46 s**

Find the elapsed time.

16. 9:35 A.M. to 11:50 A.M.
17. 7:20 A.M. to 12:55 P.M.
18. 10:30 P.M. to 6:15 A.M.
19. 8:10 A.M. to 3:05 P.M.

16. 2 h 15 min
17. 5 h 35 min
18. 7 h 45 min
19. 6 h 55 min

20. **Olympics** In the 2000 Olympics, Robert Korzeniowski of Poland won the gold medal in the 50-kilometer walk with a time of 3 hours 42 minutes 22 seconds. In the 1960 Olympics, Donald Thompson of Great Britain won the gold medal in the 50-kilometer walk with a time of 4 hours 25 minutes 30 seconds. How much longer did it take Donald Thompson to complete the 50-kilometer walk in the 1960 Olympics than Robert Korzeniowski in the 2000 Olympics? **43 min 8 s**

21. **Cooking** Latanya is making a roast that needs to cook for $3\frac{1}{2}$ hours in the oven. If she put the roast in the oven at 2:53 P.M., what time will the roast finish cooking? **6:23 P.M.**

PREREQUISITE CONCEPT 9 Geometry

What You'll Learn

You'll learn to identify and draw models of the basic figures involved in the study of geometry.

A **point** is the simplest figure studied in geometry. A point is an exact location in space with no size or shape. A point can be expressed by a dot and named with a capital letter.

A •

point *B* •

Two points and all points on a straight path between them is called a **line segment**. A line segment is named by its endpoints.

segment *CD* or $\overline{CD}$ or segment *DC* or $\overline{DC}$

If a segment connecting two points is extended indefinitely in both directions, it is called a **line**. Arrowheads are used to show that a line has no endpoints. A line can be named by two points on the line or with a single lowercase letter.

line *EF* or $\overleftrightarrow{EF}$ or line *FE* or $\overleftrightarrow{FE}$ or line ℓ

In geometry, a **ray** is defined as a segment that is extended indefinitely in one direction. A ray is named using the endpoint first and then any other point on the ray.

ray *GH* or $\overrightarrow{GH}$

Examples

Refer to the figure at the right.

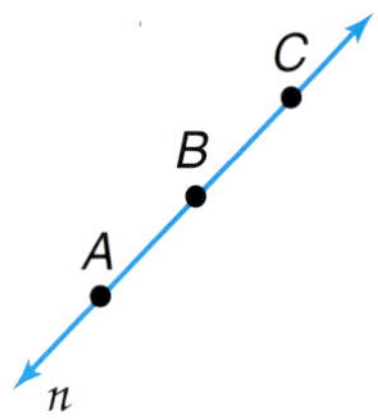

1 Name two points.

Two points are point *A* and point *B*.

2 Give three names for the line.

Any two points on the line or the lowercase letter can be used to name it. Three names are $\overleftrightarrow{AC}$, $\overleftrightarrow{BC}$, and line *n*.

3 Name two rays.

Two of the rays are $\overrightarrow{BC}$ and $\overrightarrow{BA}$.

4 Name a line segment.

One of the line segments shown is $\overline{AC}$.

When two rays have a common endpoint, they form an **angle**. The common endpoint is called the **vertex**, and the rays are called the **sides**. An angle is named using three points with the vertex as the middle letter. If no other angle shares the same vertex, the angle can be named with just the vertex letter or with a numeral.

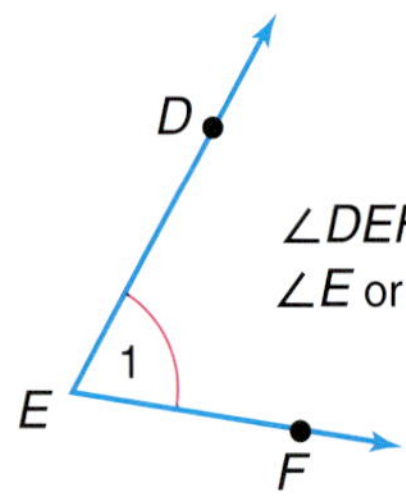

$\angle DEF$ or $\angle E$ or $\angle 1$

Example 5 **Name the angle in four ways. Then identify its vertex and its sides.**

The angle can be named in four ways: $\angle ABC$, $\angle CBA$, $\angle B$, and $\angle 1$.

Its vertex is point B. Its sides are $\overrightarrow{BA}$ and $\overrightarrow{BC}$.

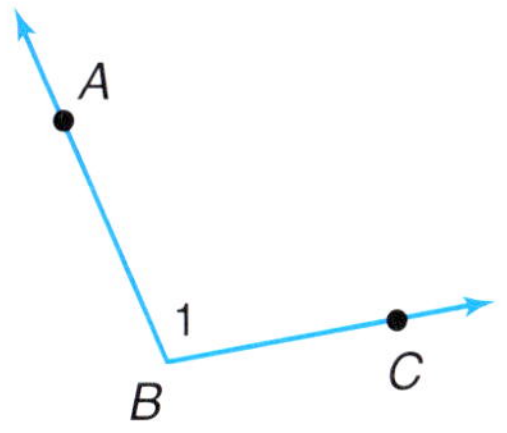

Another basic geometric figure is a **plane**. A plane is a flat surface with no edges, or boundaries. A plane can be named by a single uppercase script letter or by using three points on the plane.

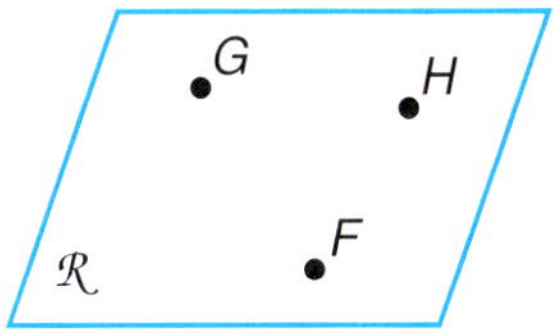

Plane *FGH* or plane $\mathcal{R}$

Use the figure at the right to name examples of each term.

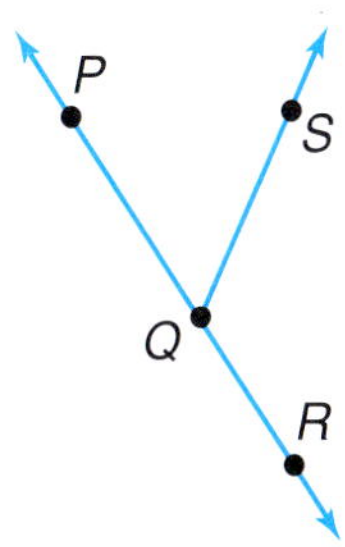

1. three points **Sample answer: *P*, *Q*, *R***
2. a line **Sample answer: $\overleftrightarrow{PR}$**
3. two rays **Sample answer: $\overrightarrow{QR}$, $\overrightarrow{QS}$**
4. two line segments **Sample answer: $\overline{PR}$, $\overline{QS}$**
5. two angles **$\angle PQS$, $\angle SQR$**
6. a point that is not on $\overleftrightarrow{PR}$ ***S***
7. an angle that does not contain point P **$\angle SQR$**

8–16. See Teacher's Answer Key.

Draw and label a diagram to represent each of the following.

8. point A
9. line b
10. plane XYZ
11. $\overline{EF}$
12. $\overrightarrow{GH}$
13. $\angle D$
14. plane $\mathcal{Q}$
15. $\angle TUV$
16. $\overleftrightarrow{RS}$

Determine whether each model suggests a *point*, a *line*, a *ray*, a *segment*, or a *plane*.

17. the tip of a pencil **a point**
18. a sheet of paper **a plane**
19. a beam from flashlight **a ray**
20. a crosswalk **a line segment**

21. **Carpentry** Mrs. Kwan wants to build some stools that won't wobble. Explain why she should make the stools with three legs. **Sample answer: The ends of the three legs determine a plane.**

PREREQUISITE CONCEPT 10

Perimeter and Area

What You'll Learn

You'll learn to find the perimeter and area of rectangles and squares, and to find the area of triangles, trapezoids, and circles.

The distance around a geometric figure is called its **perimeter**. The measure of the surface enclosed by a geometric figure is called its **area**.

Perimeter and Area of a Rectangle

The perimeter of a rectangle is twice the sum of its length ℓ and width w.

$$P = 2(\ell + w)$$

The area of a rectangle is the product of its length ℓ and width w.

$$A = \ell w$$

A square is a special rectangle in which the lengths of all the sides are equal. For this reason, the formula for the perimeter of square is $P = 4s$ and the formula for the area of a square is $A = s^2$.

Example 1 Find the perimeter and area of rectangle $ABCD$.

$P = 2(\ell + w)$

$= 2(6 + 4)$

$= 2(10)$ or 20 The perimeter is 20 inches.

$A = \ell w$

$= 6 \cdot 4$ or 24 The area is 24 square inches.

Use the formulas below to find the areas of other plane figures.

Area Formulas for Parallelograms, Triangles, Trapezoids, and Circles	Parallelograms	Triangles	Trapezoids	Circles
	The area of a parallelogram is the base times the height. $A = bh$	The area of a triangle is one-half the base times the height. $A = \frac{1}{2}bh$	The area of a trapezoid is one-half the height times the sum of the bases. $A = \frac{1}{2}h(b_1 + b_2)$	The area of a circle is π times the square of the radius. $A = \pi r^2$

Examples 2 Find the area of each figure.

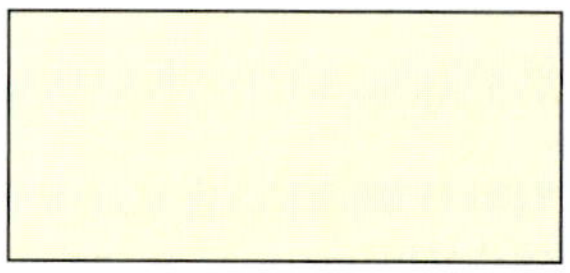

$A = bh$

$= 11 \cdot 4$ or 44

The area is 44 ft^2.

3

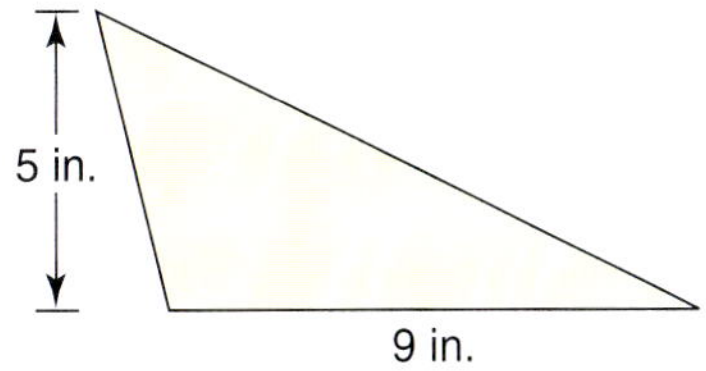

$A = \frac{1}{2}bh$

$= \frac{1}{2}(5)(9)$ or $22\frac{1}{2}$

The area is $22\frac{1}{2}$ in^2.

4

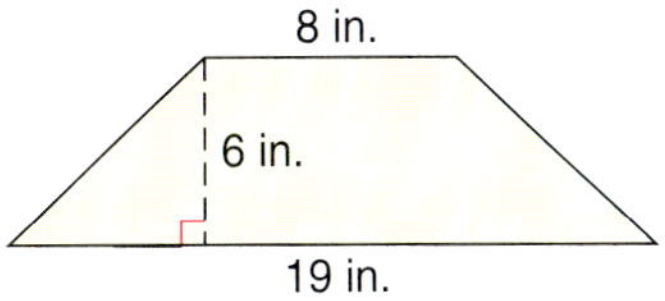

$A = \frac{1}{2}h(b_1 + b_2)$

$= \frac{1}{2} \cdot 6(8 + 19)$

$= \frac{1}{\underset{1}{2}} \cdot \frac{\overset{3}{6}}{1} \cdot 27$ or 81

The area is 81 in^2.

5

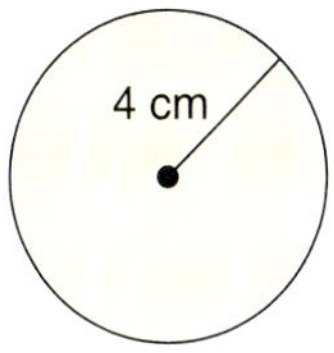

$A = \pi r^2$

$\approx 3.14(4^2)$

$\approx 3.14(16)$

≈ 50.24

The area is about 50 cm^2.

1. **Draw and label** a rectangle that has a perimeter of 12 inches.
2. **Draw and label** a triangle that has an area of 30 square centimeters.
   1–2. See Teacher's Answer Key.

Find the perimeter and area of each figure.

3. 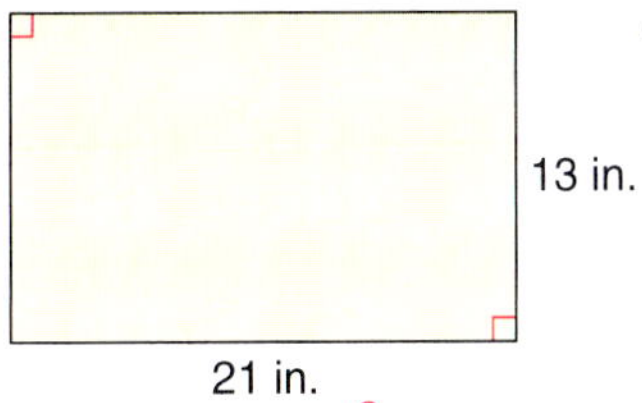

68 in.; 273 in^2

4. 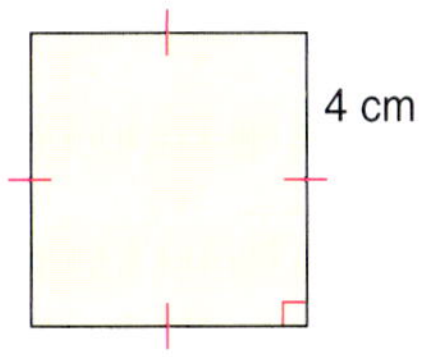

16 cm; 16 cm^2

5. 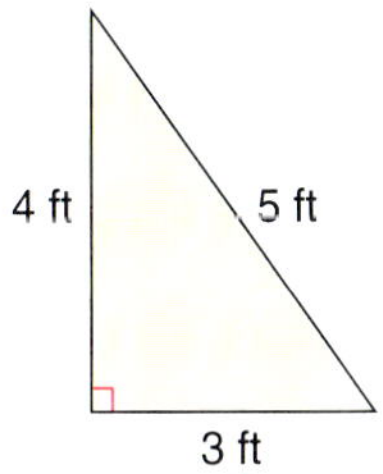

12 ft; 6 ft^2

Find the area of each figure.

6. 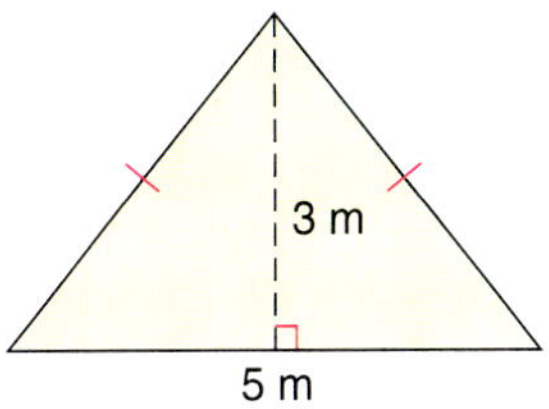

7.5 m^2

7. 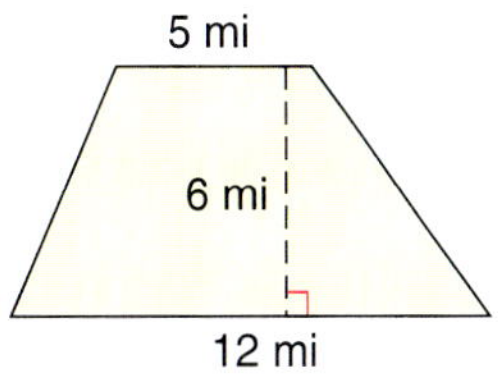

51 mi^2

8. 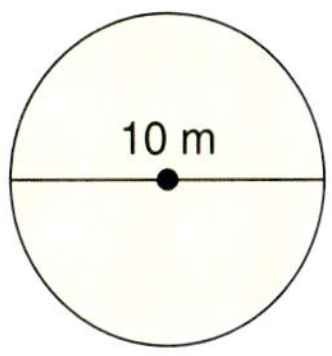

78.5 m^2

9. **Home Maintenance** How much will it cost to tile the floor of a rectangular room if the tiles cost \$1.29/ft^2 and the room is 12 ft by 18 ft? **\$278.64**

10. **Gardening** Each participant in a community vegetable garden is allotted a rectangular plot that covers 1620 ft^2. Each plot is 18 ft wide.
    a. What is the length of each plot? **90 ft**
    b. What is the perimeter of each plot? **216 ft**

Surface Area and Volume

PREREQUISITE CONCEPT 11

What You'll Learn

You'll learn to find surface area and volume of prisms.

In geometry, we study three-dimensional figures. A three-dimensional figure that has two parallel congruent sides is called a prism. A **prism** is named by the shape of its parallel bases. The figure at the right is an example of a **rectangular prism**.

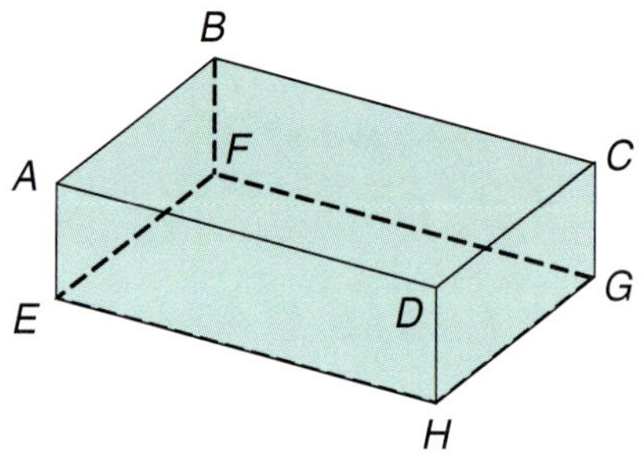

The amount of material that it would take to cover a three-dimensional figure is called the **surface area** of the figure. To find the surface area of a three-dimensional figure, find the area of each side of the figure.

Surface Area of a Rectangular Prism

Words: The surface area of a rectangular prism is the sum of the areas of the faces.

Symbols: surface area $= 2\ell w + 2\ell h + 2wh$

Model:

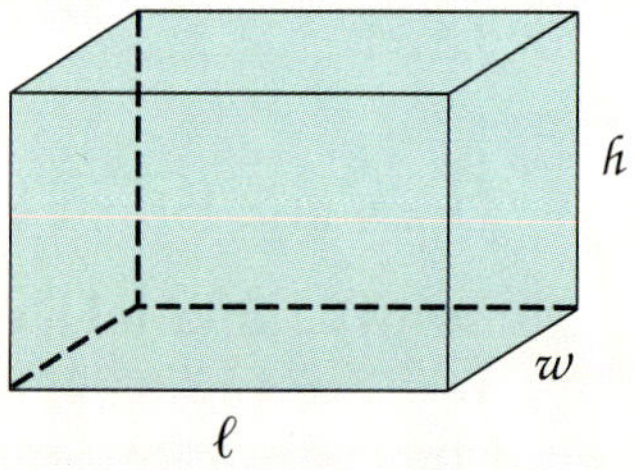

Example 1

Find the surface area of the rectangular prism at the right.

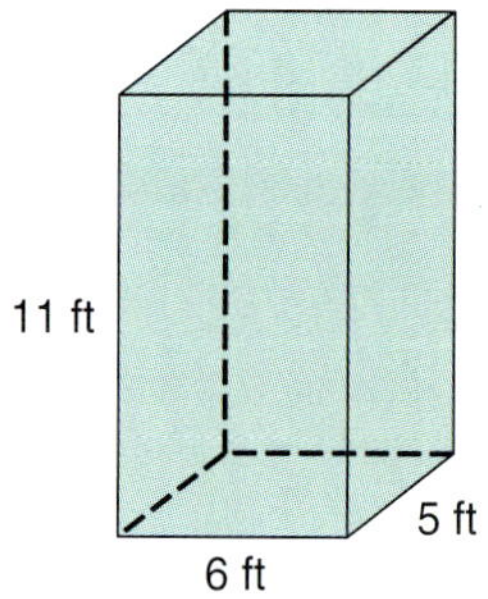

The length of the prism is 6 feet, the width is 5 feet, and the height is 11 feet.

$$\begin{aligned}\text{surface area} &= 2\ell w + 2\ell h + 2wh \\ &= 2 \times 6 \times 5 + 2 \times 6 \times 11 + 2 \times 5 \times 11 \\ &= 60 + 132 + 110 \\ &= 302\end{aligned}$$

The surface area of the rectangular prism is 302 square feet.

The amount a container will hold is called its capacity, or **volume**. Volume is usually measured in cubic units. To find the volume of a prism, multiply the area of the base times the height.

Volume of a Rectangular Prism

Words: The volume V of a rectangular prism is found by multiplying the length ℓ, the width w, and the height h.

Symbols: $V = \ell wh$

Example 2 Find the volume of the rectangular prism at the right.

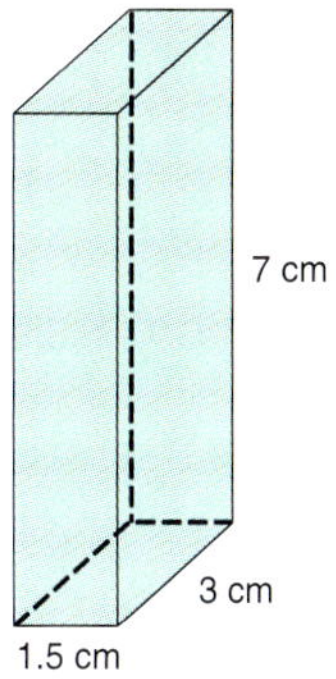

$V = \ell wh$

$= 1.5 \times 3 \times 7$

$= 31.5$

The volume of the rectangular prism is 31.5 cubic centimeters.

Find the surface area and volume of each rectangular prism to the nearest hundredth.

1.
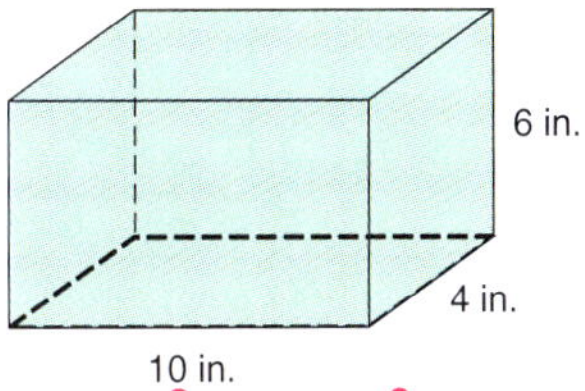

248 in²; 240 in³ → $248\ in^2$; $240\ in^3$

2.
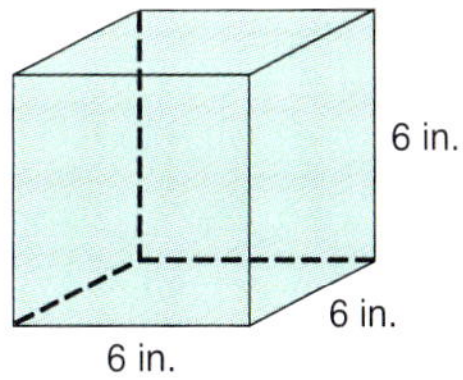

$216\ in^2$; $216\ in^3$

3.
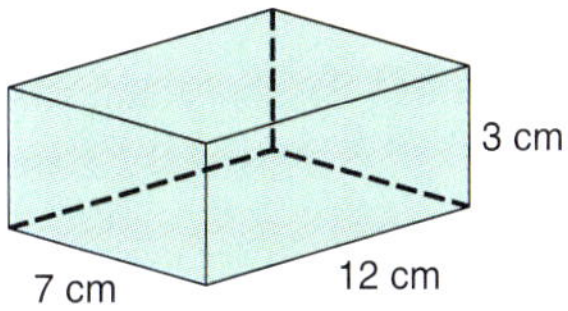

$282\ cm^2$; $252\ cm^3$

4.
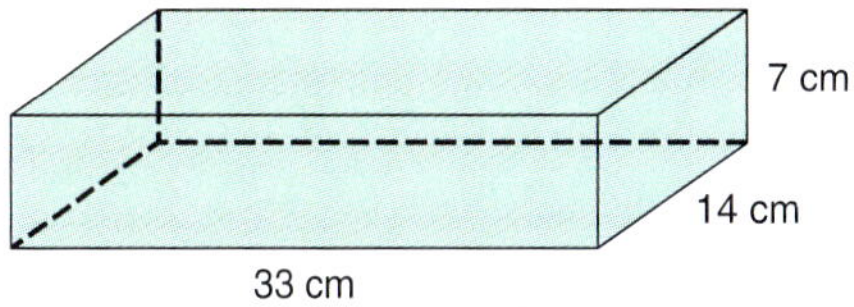

$1582\ cm^2$; $3234\ cm^3$

5.
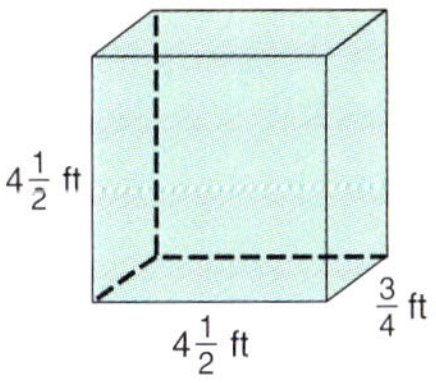

$54\ ft^2$; $15.19\ ft^3$

6.
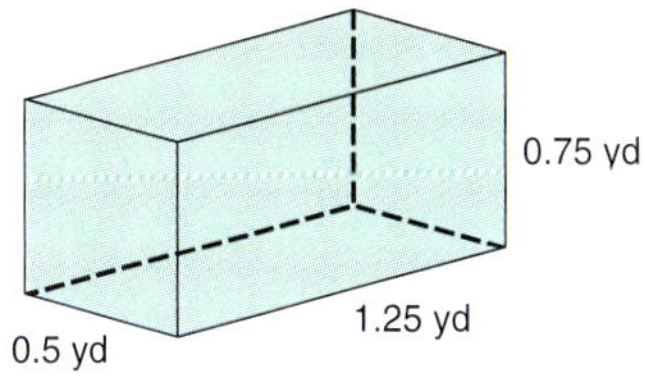

$3.88\ yd^2$; $0.47\ yd^3$

7. Explain how you can find the surface area and volume of a triangular prism.

8. Find the surface area and volume of the triangular prism at the right.

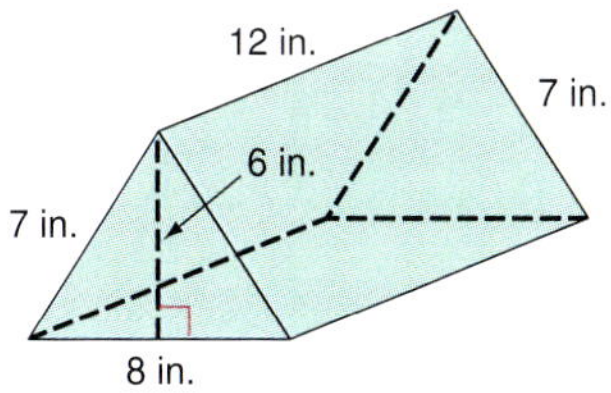

9. **Manufacturing** The case for a CD is 14.3 centimeters long, 12.4 centimeters wide, and 1.1 centimeters thick.
   a. If each CD case is to be wrapped in plastic, what is the minimum amount of plastic needed to wrap each CD? $413.38\ cm^2$
   b. How many square meters of plastic is needed to wrap 500 CDs?

10. **Pet Care** Jeremy has an end table that doubles as a fish tank. The base of the table is a rectangle that is 10 inches wide and 26 inches long. The table is 25 inches tall.
    a. What is the surface area of the table? $2320\ in^2$
    b. How many cubic feet of water does it hold? about $3.76\ ft^3$

7. To find the surface area, add the areas of each face. To find the volume, multiply the area of one triangular base by the height.

8. $312\ in^2$; $288\ in^3$

9b. about $20.67\ m^2$

Posttest

Circle the letter of the correct answer.

1. The bar graph shows the number of hours people in the U.S. work in a typical work week. What percent of those surveyed work fewer than 40 hours per week? **Source:** *USA TODAY* **C**

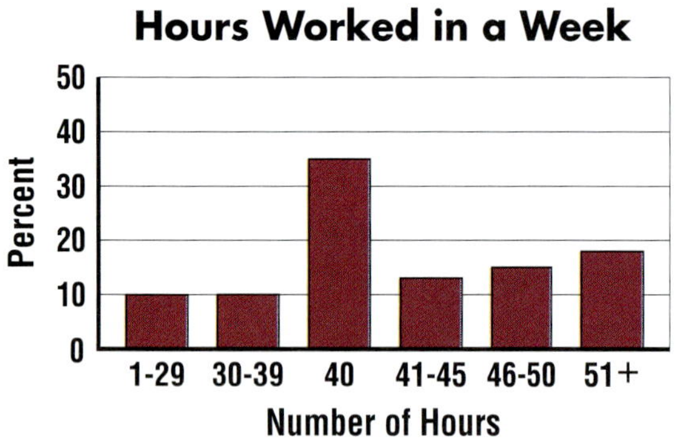

A 10%
B 15%
C 20%
D 55%

2. A card is chosen at random from the set shown below. Find the probability of choosing an even number. **G**

1	2	3	4
5	6	7	8
9	10	11	12

F $\frac{3}{4}$
G $\frac{1}{2}$
H $\frac{2}{3}$
I $\frac{1}{4}$

3. The line graph shows projected sales of digital videodisks or DVDs. Which of the following is a reasonable conclusion from the information on the graph? **Source:** Adams Media Research **B**

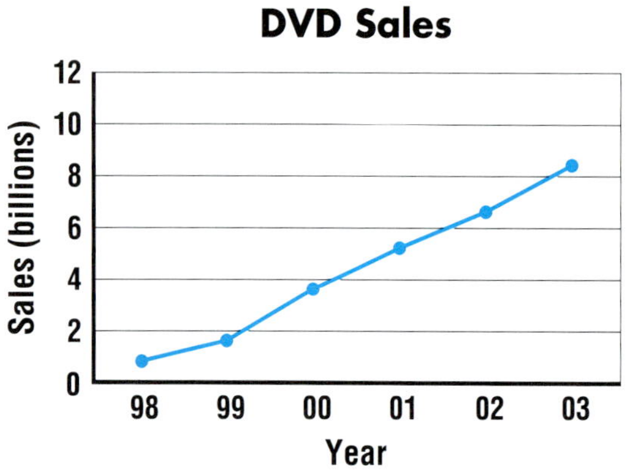

A Sales tripled between 1998 and 1999.
B Sales for 2005 are projected to be more than 10 billion dollars.
C Sales will reach their peak in 2003.
D Sales between 2002 and 2003 are projected to increase slightly.

4. Complete: 0.709 L = ______ mL. **H**
F 7.09
G 70.9
H 709
I 0.709

5. How many slices of bread, each $\frac{3}{4}$-inch thick, can be cut from a loaf of bread 12 inches long? **B**
A 14
B 16
C 12
D 10

Posttest

6. The graph shows how much money adults paid for a book. How many more of those surveyed paid \$5–\$7.99 than those who paid under \$3.00? **Source:** *USA TODAY* F

Buying Books

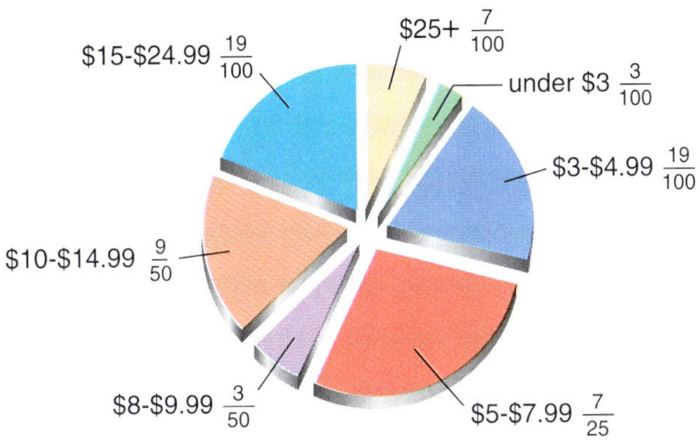

F $\frac{1}{4}$ **G** $\frac{2}{25}$

H $\frac{4}{25}$ **I** $\frac{1}{10}$

7. What is the weight in pounds of a wheel of cheese that weighs 64 ounces? C

A 5.3 lb
B 6 lb
C 4 lb
D 8 lb

8. Add 4 hours 12 minutes 39 seconds and 8 hours 6 minutes 24 seconds. H

F 13 h 19 min 3 s
G 11 h 18 min 3 s
H 12 h 19 min 3 s
I 11 h 18 min 53 s

9. How many pieces of lumber that measure $\frac{4}{5}$ foot can be cut from a piece of lumber that is $6\frac{2}{5}$ feet long? B

A 7
B 8
C 5
D 9

10. Find the perimeter of the figure. H

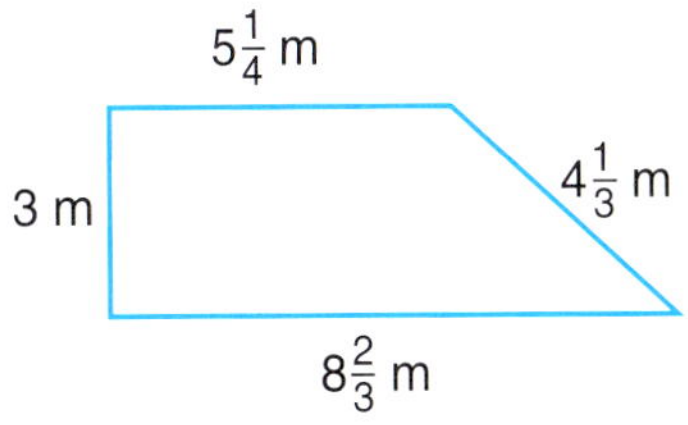

F $20\frac{2}{5}$ m **G** $20\frac{1}{4}$ m

H $21\frac{1}{4}$ m **I** $21\frac{1}{3}$ m

11. Find the product of $2\frac{1}{3}$ and $1\frac{1}{8}$. D

A $2\frac{3}{4}$ **B** $2\frac{1}{2}$

C $3\frac{1}{8}$ **D** $2\frac{5}{8}$

12. What is the area of the figure? F

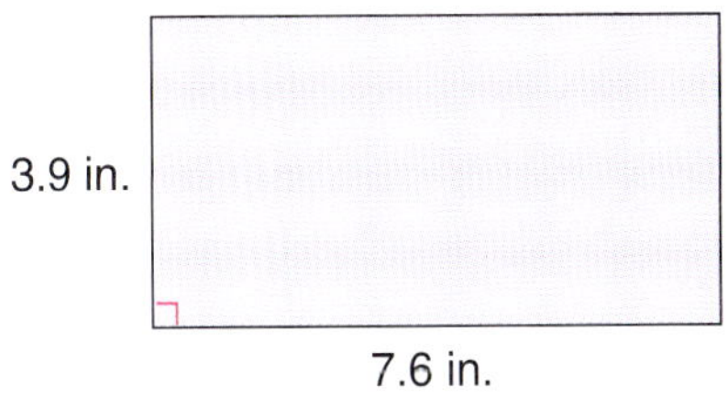

F 29.64 in^2 **G** 23.14 in^2

H 29.54 in^2 **I** 296.4 in^2

13. Sue has $7\frac{2}{3}$ cups of flour. She uses $4\frac{1}{4}$ cups of the flour to make a cake. How much flour does she have left? B

A $3\frac{1}{12}$ c **B** $3\frac{5}{12}$ c

C $2\frac{7}{12}$ c **D** $3\frac{1}{3}$ c

14. If it takes Omar 70.44 minutes to travel 12 miles on his bicycle, what is his average time per mile? F

F 5.87 min
G 6.7 min
H 6.87 min
I 7.12 min

Posttest

15. After galloping $1\frac{7}{8}$ miles, a racehorse trotted an additional $\frac{7}{8}$ miles to cool down. How far did the horse travel altogether? **D**

A $1\frac{3}{4}$ mi **B** $2\frac{1}{8}$ mi

C $2\frac{1}{4}$ mi **D** $2\frac{3}{4}$ mi

16. Margurite weighed $7\frac{1}{2}$ pounds at birth. Her brother weighed $7\frac{3}{4}$ pounds when he was born. What was the difference in their birth weights? **G**

F $\frac{1}{8}$ lb **G** $\frac{1}{4}$ lb

H $\frac{3}{8}$ lb **I** $\frac{1}{2}$ lb

17. Which figure best represents ray XY? **B**

A 

B 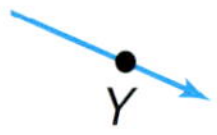

C 

D 

18. The length of 1 side of an equilateral triangle is 2.2 meters. What is its perimeter in *centimeters*? **H**

F 6.6 cm

G 66 cm

H 660 cm

I 6600 cm

19. What is the area of the trapezoid? **B**

A 10 ft^2

B 30 ft^2

C 40 ft^2

D 95 ft^2

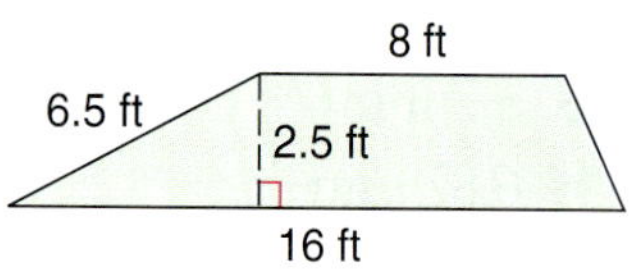

20. What is the volume of the cube? **I**

F 200 ft^3

G 175 ft^3

H 36 ft^3

I 216 ft^3

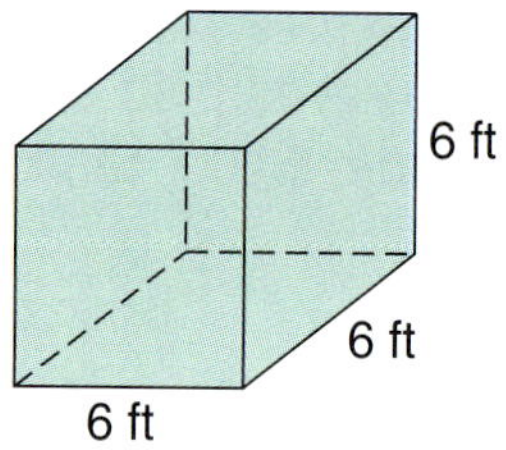

21. A bag contains 5 blue, 10 black, 5 red, and 5 yellow marbles. If you select one marble at random, what is the probability of selecting a black marble? **A**

A $\frac{2}{5}$ **B** $\frac{1}{5}$

C $\frac{2}{9}$ **D** $\frac{1}{6}$

22. Find the average number of points Jed scored in each basketball game. **G**
Jed's points: 7, 11, 12, 4, 7, 11, 19, 13, 6

F 7 **G** 10

H 11 **I** 12

23. What is the surface area of the rectangular prism? **B**

A 222 ft^2

B 546 ft^2

C 642 ft^2

D 1323 ft^2

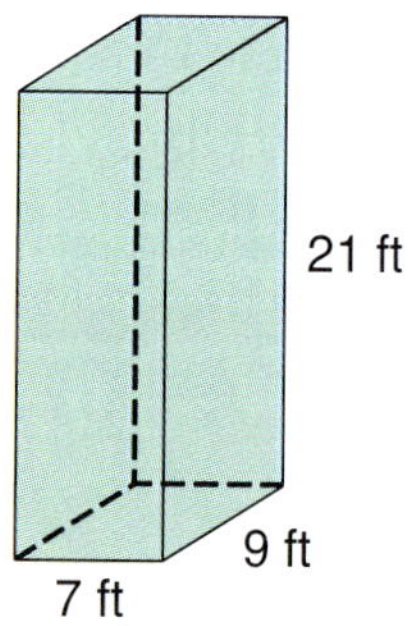

24. What is the elapsed time from 11:30 P.M. to 6:05 A.M.? **I**

F 4 hours 45 minutes

G 5 hours 25 minutes

H 6 hours 15 minutes

I 6 hours 35 minutes

25. Aliya bought 1.75 pounds of turkey. How many ounces of turkey did Aliya buy? **C**

A 10.9 ounces

B 24.5 ounces

C 28 ounces

D 175 ounces

Chapter A Teacher's Answer Key

Page A23 Prerequisite Concept 9

8. • *A*

9. 

10. 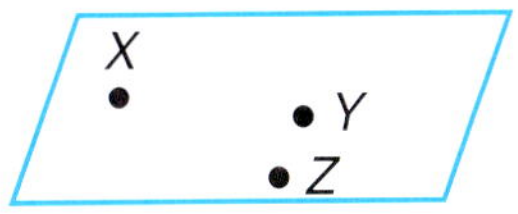

11. 

12. 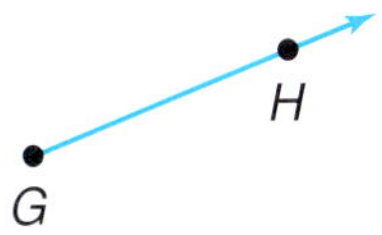

13. 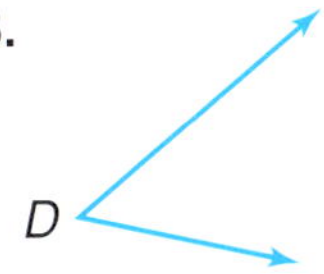

14. 

15. 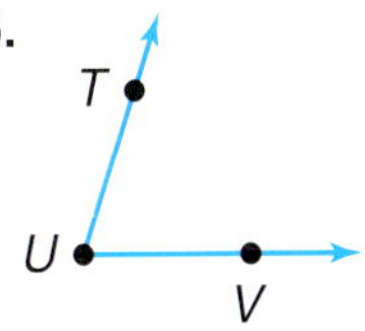

16. 

Page A25 Prerequisite Concept 10

1. 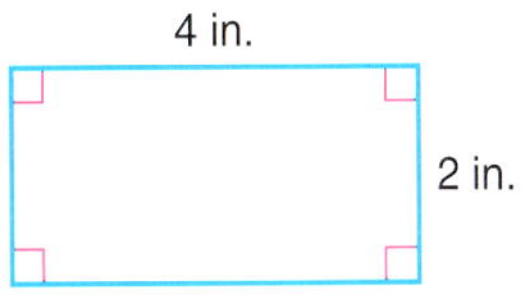

2. 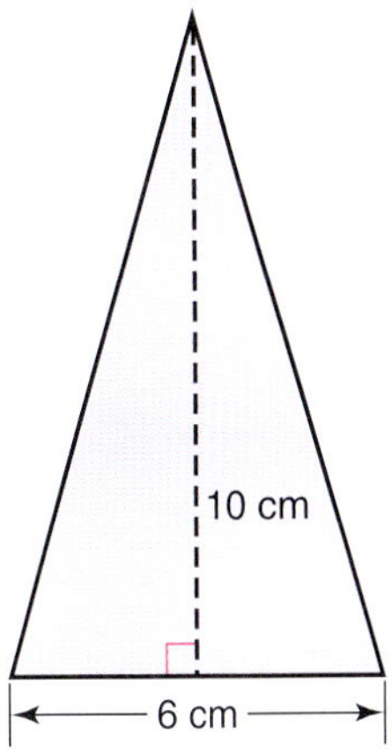

To The Student

Chapter B contains four sections: pretest, review lessons, chapter tests, and posttest. The pretest is a review of the concepts that you will need to succeed in the second half of *Algebra: Concepts and Applications.* You should take the pretest to determine which concepts you need to review. The review lessons allow you to develop, and eventually master, the individual skills the pretest identified as needing to be reinforced. You should take the posttest to make sure you understand all of the concepts and to measure your progress.

Pretest

Circle the letter of the correct answer.

1. Write the expression *two less than three times c* algebraically. **C**

   A $2 - 3c$ **B** $\frac{3c}{2}$ **C** $3c - 2$ **D** $2c - 3$

2. Find the value of x if $-17 + 8 = x$. **G**

   F -11 **G** -9 **H** 9 **I** 25

3. Simplify $-0.47 + 2.31 + (-8.07)$. **A**

   A -6.23 **B** -5.29 **C** 5.29 **D** 9.91

4. The line graph shows the number of annual visitors to Yosemite National Park. Which is a reasonable conclusion from the information on the graph? **F**

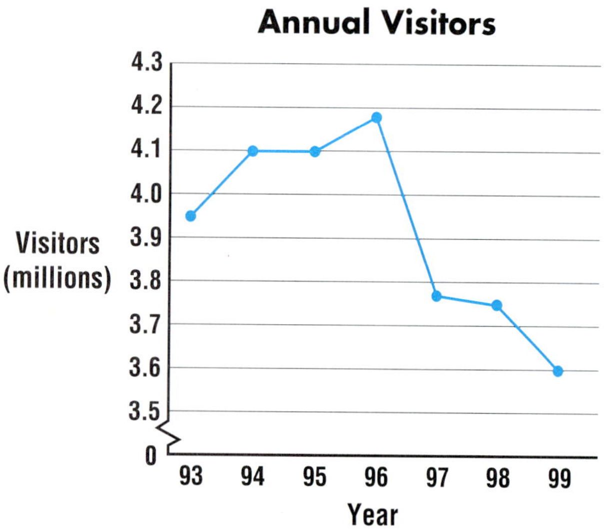

   F The average number of visitors each year is about 4 million.

   G The number of visitors increases each year.

   H More people visited the park in 1995 than in any other year.

   I Since 1997, the number of visitors has decreased by one million each year.

5. Name the property shown by $(m \cdot n) \cdot 7 = m \cdot (n \cdot 7)$. **B**

   A Associative Property of Addition

   B Associative Property of Multiplication

   C Commutative Property of Multiplication

   D Distributive Property

6. What is the ordered pair for point M? **H**

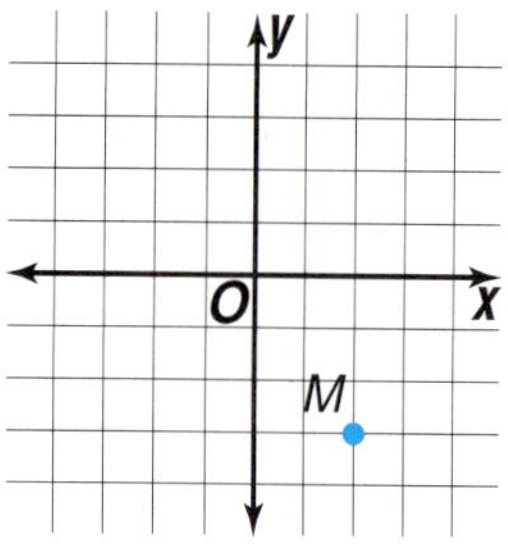

   F $(-3, 2)$ **G** $(-2, -3)$

   H $(2, -3)$ **I** $(2, 3)$

7. Find the value of $[a + 2(3 + b)] \div 4$ if $a = 8$ and $b = 5$. **A**

   A 6 **B** 8 **C** 12 **D** 20

8. Evaluate $\frac{-5y}{x}$ if $x = -3$ and $y = -9$. **G**

   F -18 **G** -15 **H** 15 **I** 18

9. What is the solution of $58 = 29 + z$? **A**

   A 29 **B** 31 **C** 39 **D** 87

10. Solve $|x - 2| + 7 = 13$. **G**

    F $\{-18, 22\}$ **G** $\{-4, 8\}$

    H $\{-8, 4\}$ **I** $\varnothing$

11. The table shows the number of acres for the six largest city-owned parks in the U.S. Find the mean size of the parks. **C**

Park	Size
Cullen (Houston)	10,534
Fairmount (Philadelphia)	8700
Griffith (Los Angeles)	4218
Eagle Creek (Indianapolis)	3800
Pelham Bay (Bronx, N.Y.)	2764
Mission Bay (San Diego)	2300

Source: Urban Land Institute/Trust for Public Land

    A 4009 acres **B** 4218 acres

    C 5386 acres **D** 8234 acres

12. What is the solution of $-\frac{1}{5}w = -45$? **I**

    F -225 **G** -9 **H** 9 **I** 225

Pretest

13. The weight of an object on Mars is $\frac{3}{8}$ of its weight on Earth. Suppose an eagle weighs $12\frac{2}{3}$ pounds on Earth. How much would it weigh on Mars? **C**

A $4\frac{2}{3}$ lb B $4\frac{5}{8}$ lb

C $4\frac{3}{4}$ lb D $33\frac{7}{9}$ lb

14. Solve $4(a - 3) - 5(3a + 6) = 13$. **H**

F 5 G −1 H −5 I −7

15. According to the scale on a map, 1 centimeter represents 20.5 kilometers. Suppose two towns are 150 kilometers apart. Which proportion could be used to find d, the distance between the two cities on the map? **A**

A $\frac{1}{20.5} = \frac{d}{150}$ B $\frac{20.5}{1} = \frac{d}{150}$

C $\frac{1}{d} = \frac{150}{20.5}$ D $\frac{150}{1} = \frac{20.5}{d}$

16. Suppose the spinner shown is spun once. Find the odds of landing on a number greater than or equal to 4. **I**

F 1:2 G 3:5 H 4:4 I 5:3

17. What is the range of the relation {(3, 4), (5, 5), (−1, 7), (6, 8), (0, −5)}? **D**

A {3, 5, −1, 6, −2, 0} B {−5, −4, 7, 5, 4}

C {4, 5, 7, 8} D {4, 5, 7, 8, −5}

18. What number is 36% of 165? **H**

F 5.94 G 58.4 H 59.4 I 63

19. In May 1997, farmers in the U.S. received an average of \$8.40 per bushel of soybeans. In December 1999, they received \$4.29 per bushel. What was the percent of decrease? Round to the nearest percent. **B**

A 47% B 49% C 51% D 53%

20. Which relation is a function? **G**

F {(5, −3), (5, 5), (0, 2), (6, 4)}

G

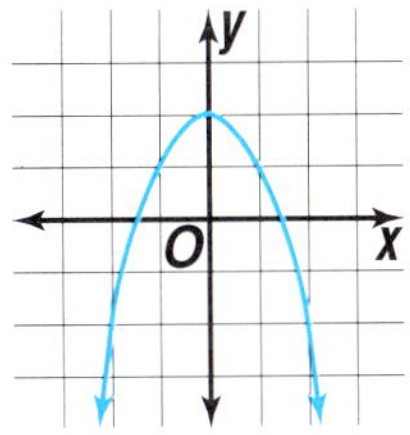

H

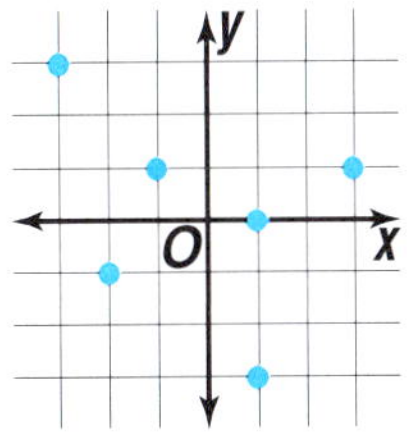

I

x	y
0	2
1	3
−5	−2
2	−3
1	4

21. Suppose y varies directly as x and $y = 18$ when $x = 9$. Find x when $y = 32$. **A**

A 16 B 12 C 8 D 6

22. What is the y-intercept of $2x - 3y = 6$? **I**

F (3, 0) G (0, 2)

H (−3, 0) I (0, −2)

23. Find the slope of the line that passes through (8, 3) and (4, 5). **A**

A $-\frac{1}{2}$ B $\frac{1}{2}$ C $\frac{2}{3}$ D 2

24. Find an equation of the line through (−8, 3) with slope $-\frac{1}{2}$. **H**

F $y = -\frac{1}{2}x + 7$ G $y = -\frac{1}{2}x + 3$

H $y = -\frac{1}{2}x - 1$ I $y = -\frac{1}{2}x + 1$

25. A line that is parallel to the graph of $4x - 5y = 8$ has which of the following slopes? **C**

A $-\frac{5}{4}$ B $-\frac{4}{5}$ C $\frac{4}{5}$ D $\frac{5}{4}$

REVIEW 1-1

Writing Expressions and Equations

Any letter used to represent an unknown number is called a **variable.** You can use variables to translate verbal expressions into **algebraic expressions.**

Words	Symbols
4 more than a number	$y + 4$
a number decreased by 12	$b - 12$
the product of 3 and a number	$3t$
a number divided by 8	$h \div 8$ or $\frac{h}{8}$

An **equation** is a sentence that contains an equals sign (=). Phrases such as *is equal to, is the same as,* and *is equivalent to* indicate the equals sign.

Example

Write an equation for *Five subtracted from n is 13.*

Five subtracted from n is 13. → $n - 5 = 13$

Exercises Write an algebraic expression for each verbal expression.

1. $n + 14$

5. $n + 17$

1. A number increased by 14
2. six times a number $6n$
3. 12 more than a number $12 + n$
4. a number divided by 4 $n \div 4$
5. the sum of a number and 17
6. 25 less than a number $n - 25$
7. three plus the product of a number and 13 $3 + 13n$
8. 74 decreased by twice a number $74 - 2n$

9–11. Sample answers are given.
9. one more than z
10. five times b
11. fifty-seven decreased by the product of 3 and q

Write a verbal expression for each algebraic expression.

9. $z + 1$
10. $5b$
11. $57 - 3q$

Write an equation for each sentence.

12. The product of 4 and x is equivalent to 12. $4x = 12$
13. Twenty divided by y is equal to 10. $20 \div y = 10$
14. Eleven is the same as the sum of a, b, and 7. $11 = a + b + 7$

15. **Environment** According to the American Water Works Association, it takes an average of 20 gallons of water to wash dishes by hand and 25 gallons of water to run a dishwasher.
    a. Write an expression that represents the amount of water used to wash dishes by hand for x washes. $20x$
    b. Write an expression that represents the amount of water needed to run a dishwasher for x washes. $25x$
    c. Write an expression that represents the difference between the amount of water needed to run a dishwasher and the amount of water used to wash dishes by hand for x washes. $25x - 20x$

REVIEW 1-2

Order of Operations

Ian said Shelley won first prize and I won second prize. Without punctuation, this sentence has three possible meanings.

Ian said, "Shelley won first prize and I won second prize."

"Ian," said Shelley, "won first prize and I won second prize."

Ian said Shelley won first prize and I [that is, the speaker] won second prize.

In mathematics, in order to avoid confusion about meaning, an agreed-upon **order of operations** tells us whether a mathematical expression such as $45 - 15 \div 5$ means $(45 - 15) \div 5$ or $45 - (15 \div 5)$.

Order of Operations

1. Find the values of expressions inside grouping symbols, such as parentheses ( ), brackets [ ], and as indicated by fraction bars.
2. Do all multiplications and/or divisions from left to right.
3. Do all additions and/or subtractions from left to right.

You can **evaluate** an algebraic expression when the value of each variable is known. Replace each variable with its known value and then use the order of operations. Remember to do all operations within grouping symbols first.

Examples

1 **Simplify $45 - 15 \div 5$.**

$$45 - 15 \div 5 = 45 - 3$$
$$= 42$$

2 **Evaluate $a - (b + 7)$ if $a = 10$ and $b = 2$.**

$$a - (b + 7) = 10 - (2 + 7)$$
$$= 10 - 9 \text{ or } 1$$

Exercises Find the value of each expression.

1. $18 - 2 \cdot 3$ **12**
2. $16 \div 4 - 3 + 4 \cdot 6$ **25**
3. $5(36 - 31) + 2 \cdot 7$ **39**
4. $\dfrac{70 - 25}{7 + 8}$ **3**

Evaluate each algebraic expression if $a = 4$, $b = 2$, $x = 1$, and $y = 6$.

5. $8a - 2b$ **28**
6. $10x + 9y$ **64**
7. $48x - (3a - 5b)$ **46**
8. $\dfrac{a + y}{5b}$ **1**

9. **Accounting** Keisha and Mark are selling tickets for a dinner theater performance. Combination tickets for dinner and the performance cost \$15. Performance only tickets cost \$9.
   a. If Keisha and Mark sell both types of tickets, write an expression for the total amount of money they collect. **$15(d + j) + 9(d + j)$**
   b. Keisha sells 20 combination tickets and 15 performance tickets. Mark sells 10 combination tickets and 25 performance tickets. How much money have they collected? **\$810**

REVIEW 1-3

Commutative and Associative Properties

The Commutative and Associative Properties can be used to simplify algebraic expressions. To **simplify** an expression, eliminate all parentheses first and then add, subtract, multiply, or divide.

Commulative Properties of Addition and Multiplication	For any numbers a and b, $a + b = b + a$ and $a \cdot b = b \cdot a$.
Associative Properties of Addition and Multiplication	For any numbers a, b, and c, $(a + b) + c = a + (b + c)$ and $(a \cdot b) \cdot c = a \cdot (b \cdot c)$

Example

Simplify $(3 + 5n) + 12$. Identify the properties used in each step.

$(3 + 5n) + 12 = (5n + 3) + 12$ *Commutative Property of Addition*

$= 5n + (3 + 12)$ *Associative Property of Addition*

$= 5n + 15$ *Substitution Property*

Exercises Name the property shown by each statement.

1. $13a + 4b = 4b + 13a$ **Comm. (+)**
2. $1 \cdot z = z \cdot 1$ **Comm. (×)**
3. $2(x \cdot 7) = (2 \cdot x)7$ **Assoc. (×)**
4. $(5w + 7) + 6z = 5w + (7 + 6z)$ **Assoc. (+)**

Simplify each expression.

5. $5 + 6b + 12$ **$6b + 17$**
6. $(3m + 4) + 9$ **$3m + 13$**
7. $(3q + 2) + (1 + 5)$ **$3q + 8$**
8. $8 + (9y + 3)$ **$9y + 11$**
9. $(6st + 9) + 7$ **$6st + 16$**
10. $(3x + 5) + 10$ **$3x + 15$**
11. $6 \times (4 \times s)$ **$24s$**
12. $2(w) \cdot 4$ **$8w$**
13. $8 \cdot v \cdot 3$ **$24v$**
14. $(4r) \cdot 3$ **$12r$**
15. $(2q \cdot 4) \cdot 4$ **$32q$**
16. $(2 \cdot 12n) \cdot 3$ **$72n$**

17. **Automobiles** Consider the steps necessary to fill a car's gasoline tank. One step is to remove the gasoline tank cap and the other step is to pump the gasoline. Would you say that these steps are commutative? Explain. **No, the order of these steps cannot be reversed.**

REVIEW 1-4

Distributive Property

To find the product of two integers, you can find the sum of two partial products. For example, write

$$\begin{array}{r} 63 \\ \times\ 7 \\ \hline 441 \end{array} \qquad \text{as} \qquad \begin{array}{r} 60 + 3 \\ \times \qquad 7 \\ \hline 420 + 21 \end{array} \quad \leftarrow (60 \times 7) + (3 \times 7).$$

The statement $(60 + 3) \times 7 = (60 \times 7) + (3 \times 7)$ illustrates the **Distributive Property**. The multiplier 7 is distributed over the addition of 60 and 3.

Distributive Property	For any numbers a, b, and c, $a(b + c)$ 5 $ab + ac$ and $a(b - c) = ab - ac$.

You can use the Distributive Property to simplify algebraic expressions.

Example

Simplify $6(x + z) + 9z$.

$6(x + z) + 9z = 6x + 6z + 9z$	*Distributive Property*
$= 6x + (6 + 9)z$	*Distributive Property*
$= 6x + 15z$	*Substitution Property*

Name the coefficient of each term. Then name the like terms in each list of terms.

1. $12r, 7r, 3r, 4rs$
2. $3xy, 11x, 3y, 5x$

1. 12, 7, 3, 4; $12r$, $7r$, $3r$
2. 3, 11, 3, 5; $11x$, $5x$

Use the Distributive Property to rewrite each expression.

3. $7(5w + 3)$ **$35w + 21$**
4. $5a - 5b$ **$5(a - b)$**

Simplify each expression, if possible. If not possible, write in simplest form.

5. $12c - 5c$ **$7c$**
6. $4b + 9b$ **$13b$**
7. $5p + 3 - p$ **$4p + 3$**
8. $6(4x + 9)$
9. $2t + 15t + 16y - 9y$
10. $26rs + 17st$
11. $3(1 - x) - 2$
12. $2w - 3(4v + 3v)$
13. $9(3z - 2) - 7z$
14. $9(a + b) + 6b$
15. $12x - y + 4$
16. $5(2g + 1) + 6(4g - h)$

8. $24x + 54$
9. $17t + 7y$
10. in simplest form
11. $1 + 3x$
12. $2w - 21v$
13. $20z - 18$
14. $9a + 15b$
15. in simplest form
16. $34g - 6h + 5$

17. **Business** Each month, a business owner pays \$23 for telephone service and \$9.95 for an on-line service.
   a. Write an expression representing the owner's total costs for telephone and on-line services for an entire year. **12(23 + 9.95)**
   b. What amount does the business owner pay for these services for a year? **\$395.40**

REVIEW 1-5

A Plan for Problem Solving

A **formula** is an equation that states a rule for the relationship between quantities. Using a formula is one way to solve problems. However, any problem can be solved using a problem-solving plan.

1. **Explore** Read the problem carefully. Identify all information given and determine what you are asked to find.

2. **Plan** Select a strategy for solving the problem. Some strategies are shown in the box below.

Look for a pattern.	Use an equation or formula.
Draw a diagram.	Make a graph.
Make a table.	Guess and check.
Work backward.	

3. **Solve** Use your strategy to solve the problem. You may have to choose a variable for the unknown, and then write an expression. Be sure to answer the question asked.

4. **Examine** Check your answer. Does it make sense?

Exercises Solve.

1. **Geometry** The perimeter P of a square is four times the length of a side s.

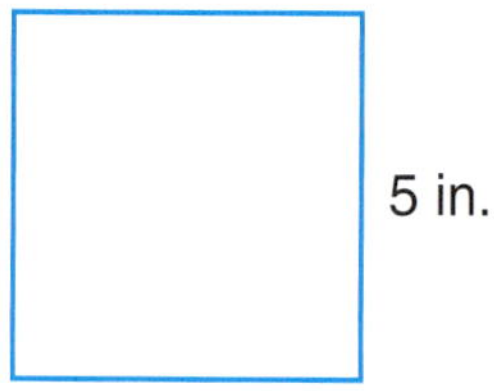

   a. Write a formula for the perimeter of a square.
   p = 4s

   b. What is the perimeter of the square shown?
   20 in.

2. **Savings** An Internet service offers two different payment plans. Plan A offers unlimited use for \$19.95 a month. Plan B offers 20 free hours a month, but then \$2.50 for each additional hour.

   a. If you use the Internet, on average, 30 hours a month, which plan is the better deal? Explain. With Plan A, you will pay \$19.95. With Plan B, you will pay 2.50(30 − 20) or \$25. So, Plan A is better.

   b. What is the maximum number of monthly hours that you can use the Internet where the better deal is with Plan B?

2b. After a maximum of 7 additional hours, Plan B is still the better deal, costing \$17.50. When there is an additional 8 hours, the cost jumps to \$20. So, Plan B is the better deal for up to 20 + 7 or 27 monthly hours.

REVIEW 1-6 Collecting Data

Sampling is a method used to gather data in which a small group, or **sample**, is used to represent a much larger **population**. Three important characteristics of a good sample are listed below.

Sampling Criteria
A good sample is: • representative of the larger population, • selected at random, and • large enough to provide accurate data.

A survey can be biased and give false results if these criteria are not followed.

Example

Twenty random people leaving a school board meeting were asked whether or not they would vote for the upcoming school levy. Eighteen or 90% of those sampled said they would vote for the levy. Is this a good sample? Explain.

No, this is not a good sample since the people leaving a school board meeting are more likely than the average person to be active in school issues and would vote for a school levy. It does not represent the larger population.

Determine whether each is a good sample. Describe what caused the bias in each poor sample. Explain.

1. The first eight people leaving a movie are asked to rate the movie on a scale from one to ten, with ten being the highest rating. **No, a smaple of eight is small. Also, the first people to leave the movie may not have liked it enough to stay until the end.**
2. All people leaving a shopping center are asked to name their favorite brand of clothing. **No, different shopping centers carry different name brands. The people leaving a certain shopping center may favor the brands that are found specifically in that center.**
3. Every other student on school files is surveyed to determine how many students can be expected to attend the next school dance. **Yes, the sample is random and would accurately represent the larger population.**
4. Two hundred randomly chosen cars of a certain make are used to determine the average number of miles per gallon of gas that make of car gets on the freeway. **Yes, the sample is random and would accurately represent the larger population.**

REVIEW **1-7**

Displaying and Interpreting Data

Mrs. Cortez interviewed some students who were willing to help her with some typing. The number of words typed per minute by each student is listed below.

16 21 18 17 18 20 15 16

These data can be organized in a **stem-and-leaf plot**.

The greatest common place value for each data item is used to form the *stem*.

Stem	Leaf
1	5 6 6 7 8 8
2	0 1

1 | 7 = 17 words per minutes

The *leaves* are formed by the next greatest place value.

In this case, the tens digits are the stems. The ones digits are the leaves. Write the leaves in order from least to greatest.

A *key* is always included. This shows how the digits are related.

In the stem-and-leaf plot at the right, the data are represented by three-digit numbers. In this case, use the digits in the first two place values to form the stems. For example, the values for 102, 114, 115, 108, 139, 127, 125, and 131 are shown in the stem-and-leaf plot at the right.

Stem	Leaf
10	2 8
11	4 5
12	5 7
13	1 9

11 | 5 = 115

Exercises Solve.

1. The stem-and-leaf plot at the right shows the height, in feet, of buildings in Boston that are at least 500 feet tall.

Stem	Leaf
5	0 0 1 2 3 5
6	0 0 0 0 1
7	5
8	0

5 | 2 = 520 feet

   a. How tall is the tallest building? **800 feet**

   b. What is the height of the shortest building represented in the plot? **500 feet**

   c. What building height occurs most frequently? **600 feet**

2. Danielle is a sales clerk at a bagel shop. Her sales for the first week were $58, $64, $26, $79, and $55. Her sales for the second week were $39, $72, $64, $52, and $78.

   a. Make a stem-and-leaf plot of Danielle's two weeks' sales.

   b. What is the greatest value of sales that Danielle made in one day during the two-week period? **$79**

2a. See Teacher's Answer Key.

CHAPTER 1 Test

1. Write an equation for the sentence *a number x subtracted from 16 is equivalent to 2.* $16 - x = 2$
2. Write an algebraic expression for *the sum of 3 times a number and 7.* $3n + 7$

Find the value of each expression.

3. $15 - 30 \div 3 + 6$ 11
4. $4[(20 - 4) \div 8]$ 8

Simplify each expression.

5. $6c + 4g - 2c$ $4c + 4g$
6. $3(7q + 1)$ $21q + 3$
7. $2 \cdot b \cdot 5$ $10b$
8. $9 + (r + 4)$ $13 + r$
9. $10h + 6(2 + h)$ $16h + 12$
10. $7y + (5y - 8)$ $12y - 8$

Name the property shown by each statement.

11. $4 + (1 + 6) = (4 + 1) + 6$ Assoc. (+)
12. $9t - 3t = t(9 - 3)$ Dist.
13. $x \cdot y = y \cdot x$ Comm. (×)

Evaluate each expression if $p = 3$, $r = 5$, and $s = 2$.

14. $s + 4r - 5$ 17
15. $r(p - s)$ 5

Stem	Leaf
7	0 4 6 7
8	1 2 9
9	3 3 5 8

8 | 1 = 81

16. Write the members of the data set used to make the stem-and-leaf plot. 70, 74, 76, 77, 81, 82, 89, 93, 93, 95, 98
17. Find the difference of the highest and lowest values of the stem-and-leaf plot. $98 - 70 = 28$
18. Which number occurs most often in the stem-and-leaf plot? 93
19. Every person in a British literature class is asked to name the average number of books read in a month to determine how literate the students in a school are. Is this a good sample? Explain. See Teacher's Answer Key.
20. **Savings** You can use the formula $I = prt$ to find the interest earned I after depositing an amount p into an account with an interest rate r after so many years t. If you deposit $100 into an account with an interest rate of 5%, after how many years will you have earned $20? 4 years

REVIEW 2-1

Graphing Integers on a Number Line

The figure at the right is part of a number line. On a number line, the set of numbers that include 0 and numbers to the right of 0 are called the set of **whole numbers**.

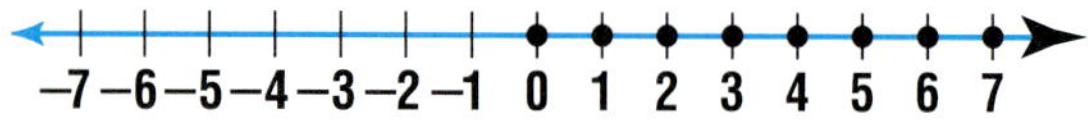

The set of numbers used to name the points marked on the number line at the right is called the set of **integers**.

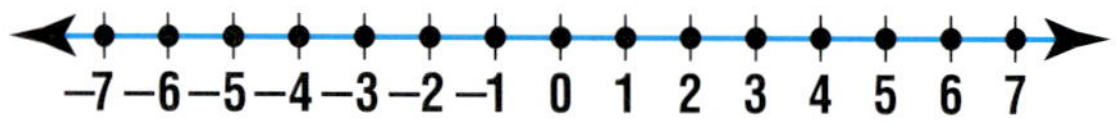

To graph a set of numbers means to locate the points named by those numbers on the number line. The number that corresponds to a point on the number line is called the **coordinate** of the point.

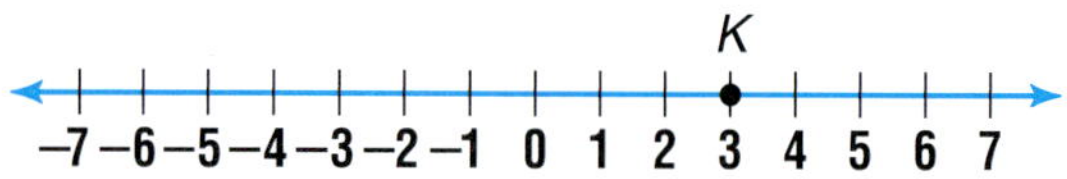

The coordinate of K is 3.

The numbers on a number line increase as you move to the right and decrease as you move to the left. When two integers are graphed on a number line, the number to the right is always greater. The number to the left is always less than the other.

Name the coordinate of each point.

1. D **2** **2.** B **–7** **3.** E **8** **4.** C **–3** **5.** A **–11** **6.** F **12**

Graph each set of numbers on a number line.

7–11. See Teacher's Answer Key

7. $\{-4, -2, 0, 3\}$ **8.** $\{-5, -1, 2, 5\}$ **9.** $\{-2, -1, 0, \ldots\}$

10. {integers that are also whole numbers}

11. {integers that are a multiple of 2}

Replace each ● with $<$ or $>$ to make a true sentence.

12. -3 ● -9 $>$ **13.** -5 ● 7 $<$ **14.** 12 ● 14 $<$

15. -3 ● 3 $<$ **16.** -8 ● -4 $<$ **17.** -1 ● -2 $>$

18. **Temperature** With the wind chill factor considered, a temperature of 8°F felt like -5°F. The next day, a temperature of 10°F felt like -11°F. On which day, the first or second, did it feel colder? **The second day**

REVIEW 2-2 The Coordinate Plane

In the diagram at the right, the two perpendicular lines, called the **x-axis** and the **y-axis,** divide the coordinate plane into four **quadrants.** The point where the two axes intersect is called the **origin.** The origin is represented by the **ordered pair** (0, 0).

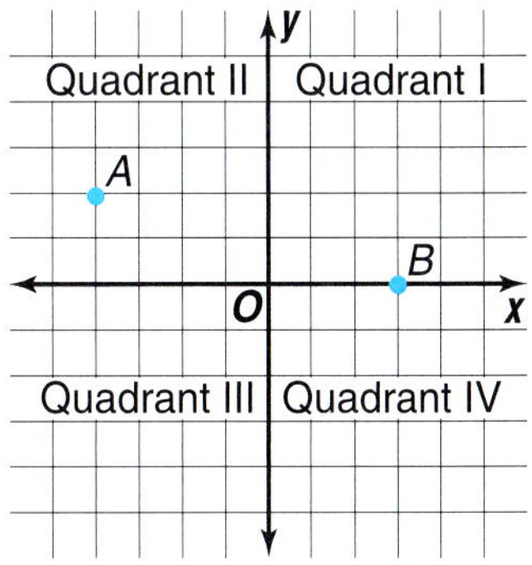

Every other point in the coordinate plane is also represented by an ordered pair of numbers. The ordered pair for point A is $(-4, 2)$. We say that -4 is the **x-coordinate** of A and 2 is the **y-coordinate** of A.

Example

Refer to the diagram above. What ordered pair names point *B*?

The x-coordinate is 3 and the y-coordinate is 0. Thus, the ordered pair is (3, 0).

To graph any ordered pair (x, y), begin at the origin. Move left or right x units, and then up or down y units. Draw a dot at that point.

Graph each point on a coordinate plane.

1–6. See Teacher's Answer Key.

1. $C(-4, 0)$
2. $D(2, 2)$
3. $E(3, -5)$
4. $F(-2, 4)$
5. $G(-1, -3)$
6. $H(0, -3)$

Write the ordered pair that names each point shown at the right. Name the quadrant in which the point is located.

7. I (1, 2); I
8. J (3, −4); IV
9. K (−2, 3); II
10. L (−5, −2); III
11. M (0, 1); none

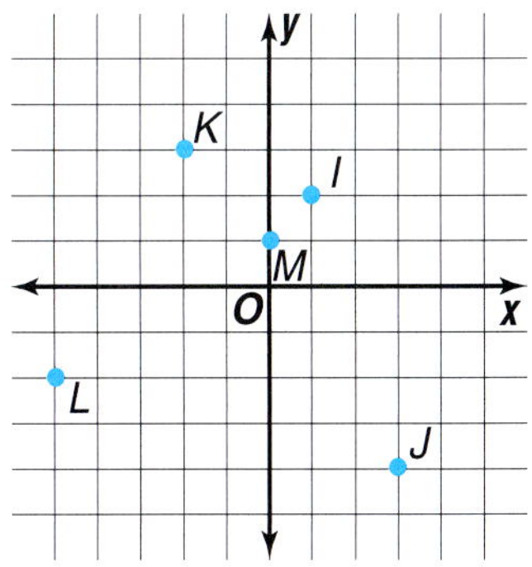

12. **Computers** Juan is a computer programmer. He is developing multimedia math software. Each screen in the program must be mapped out pixel by pixel. In the screen shown at the right, find the coordinates of each pixel making up the variable X. (3, 4), (3, 5), (3, 7), (3, 8), (4, 6), (5, 4), (5, 5), (5, 7), (5, 8)

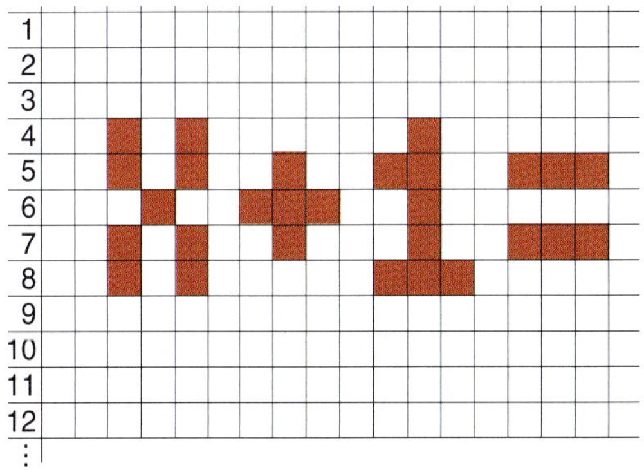

REVIEW 2-3 Adding Integers

Use the following definitions, rules, and properties when adding integers.

Definition, Rule, or Property		Numbers
Definition of Absolute Value	The absolute value of a number is the distance it is from 0 on the number line.	$\|6\| = 6$ $\|-6\| = 6$
Adding Integers with the Same Sign	To add integers with the same sign, add their absolute values. Give the result the same sign as the integers.	$3 + 2 = 5$ $-3 + (-2) = -5$
Additive Inverse Property	The sum of any number and its additive inverse is 0.	$3 + (-3) = 0$ $-7 + 7 = 0$
Adding Integers with Different Signs	To add integers with different signs, find the difference of their absolute values. Give the result the same sign as the integer with the greater absolute value.	$3 + (-2) = 1$ $-3 + 2 = -1$

You can use the Distributive Property and the rules for adding integers to simplify expressions with like terms.

Example

Simplify $-3w + (-2w)$.

$$\begin{aligned} -3w + (-2w) &= [-3 + (-2)]w \\ &= (-5)w \\ &= -5w \end{aligned}$$

Find each sum.

1. $-13 + (-7)$ **−20**
2. $96 + (-54)$ **42**
3. $39 + 62$ **101**
4. $18 + (-3)$ **15**
5. $26 + (-40)$ **−14**
6. $-27 + (-1)$ **−28**
7. $12a + (-17a)$ **−5a**
8. $-2t + 20t$ **18t**
9. $-6x + 15x$ **9x**

Evaluate each expression if $a = 2$, $b = -3$, and $c = -1$.

10. $212 + |b|$ **215**
11. $(-23) + |c|$ **−22**
12. $-34 + a$ **−32**
13. $16 + (-a)$ **14**

14. **Elevators** The Sears Tower in Chicago has 110 stories. If an employee rode the elevator to her office on the 83rd floor, then went up 15 floors to deliver a report, which floor is she on now? **98th**

REVIEW 2-4 Subtracting Integers

Use the following rule to subtract integers.

Rule	Words	Numbers
Subtracting Integers	To subtract an integer, add its additive inverse.	$13 - (-8) = 13 + 8$ or 21 $4 - 3 = 4 + (-3)$ or 1

Example

Evaluate $a - b$ if $a = -9$ and $b = 5$.

$a - b = -9 - 5$ *Replace a with −9 and b with 5.*

$= -9 + (-5)$ *Write 9 − 5 as 9 + (−5).*

$= -14$ *−9 + (−5) = −14*

Find each difference.

1. $11 - 2$ **9**
2. $7 - 4$
3. $-5 - 5$ **−10**
4. $3 - (-1)$ **4**
5. $12 - (-8)$ **20**
6. $-14 - 7$ **−21**
7. $-2 - (-6)$ **4**
8. $17 - 9$ **8**
9. $0 - 3$ **−3**
10. $12 - (-15)$ **27**
11. $25 - 10$ **15**
12. $-4 - 13$ **−17**

Evaluate each expression if $g = 6$, $h = -2$, and $k = 2$.

13. $g - h$ **8**
14. $h - k$ **−4**
15. $g - k$ **4**
16. $h - h$ **0**
17. $k - h + (-3)$ **1**
18. $5 - g + h$ **−3**

Simplify each expression.

19. $8y - 7y$ **y**
20. $6b - (-8b)$ **$14b$**
21. $-3v - v$ **$-4v$**
22. $4x - 5x$ **$-x$**
23. $-4r - 7r - 6r$ **$-17r$**
24. $9f - 5f - (-2f)$ **$6f$**

25. **Budgeting** Terri spent \$22 of the \$35 she had saved to buy a video game. How much more money would she have to save to buy a video game that costs the same price as the first one she bought? **\$9**

REVIEW 2-5

Multiplying Integers

You can use the following rules when multiplying integers.

Rule	Words	Numbers
Multiplying Two Integers with Different Signs	The product of two integers with different signs is negative.	$(-3)9 = -27$ $9(-3) = -27$
Multiplying Two Integers with the Same Sign	The product of two integers with the same sign is positive.	$-11(-4) = 44$ $11(4) = 44$

To find the product of three or more numbers, multiply the first two numbers. Then multiply the result by the next number, until you come to the end.

Example

Find $-3(7)(-1)(-5)$.

$$
\begin{aligned}
-3(7)(-1)(-5) &= -21(-1)(-5) && -3(7) = -21 \\
&= 21(-5) && -21(-1) = 21 \\
&= -105 && 21(-5) = -105
\end{aligned}
$$

Find each product.

1. $(-8)(-3)$ **24**
2. $(10)(-4)$ **−40**
3. $(-16)(2)$ **−32**
4. $-5(0)$ **0**
5. $3(3)$ 9
6. $-6(-4)$ **24**
7. $(-1)(-5)(-3)$ **−15**
8. $(-4)(-5)(-3)$ **−60**
9. $(5)(-2)(-1)(-3)$ **−30**

Evaluate each expression if $x = 4$, $y = -1$, and $z = -2$.

10. $-8y$ **8**
11. $5yz$ **10**
12. $3x + 2y$ **10**
13. $xyz + x$ **12**

14. **Nutrition** The recommended daily allowance (RDA) of calcium for a teenaged female is 1200 mg. An eight-ounce serving of skim milk contains 302 mg of calcium. A one-ounce serving of cheddar cheese contains 204 mg of calcium. A medium orange contains 50 mg of calcium. **a. 906 mg**
    a. How much calcium is in 3 eight-ounce servings of skim milk?
    b. Do 3 eight-ounce servings of skim milk satisfy the RDA of calcium for a teenaged female? Explain your reasoning. **No, 906 < 1200.**
    c. Would a serving of cheddar cheese and 2 oranges, in addition to 3 servings of skim milk, satisfy the RDA of calcium for a teenaged female? Explain your reasoning. **Yes, 204 + 2(50) + 906 = 1210, which is greater than 1200.**

REVIEW 2-6

Dividing Integers

Use the following rules when dividing integers.

Rule	Words	Numbers
Dividing Two Integers with the Same Sign	The quotient of two integers with the same sign is positive. $-40 \div (-8) = 5$	$40 \div 8 = 5$
Dividing Two Integers with Different Signs	The quotient of two integers with different signs is negative. $63 \div (-9) = -7$	$-63 \div 9 = -7$

Recall that fractions are another way of showing division. So, you can use the division rules above to evaluate rational expressions.

Example

Evaluate $\frac{-b}{4a}$ if $b = -16$ and $a = -2$.

$\frac{-b}{4a} = \frac{-(-16)}{4(-2)}$ *Replace b with −16 and a with −2.*

$= \frac{16}{-8}$ *$-(-16) = 16$ and $4(-2) = -8$*

$= -2$ *$\frac{16}{-8}$ means $16 \div -8$.*

Find each quotient.

1. $49 \div (-7)$ **−7**
2. $-36 \div 9$ **−4**
3. $18 \div 2$ **9**
4. $-24 \div (-3)$ **8**
5. $30 \div (-6)$ **−5**
6. $-15 \div (-5)$ **3**
7. $32 \div 8$ **4**
8. $-64 \div 8$ **−8**
9. $-42 \div (-6)$ **7**

Evaluate each expression if $x = 3$, $y = -1$, and $z = -12$.

10. $\frac{z}{x}$ **−4**
11. $\frac{8 - x}{y}$ **−5**
12. $\frac{yz}{x}$ **4**

13. **Parties** Caitlyn brought in a bag of 45 chocolates to share with the class for her birthday. She would like to distribute the chocolates evenly among her classmates.

    a. If there are 15 students in the class, including herself, how many chocolates does each person get? **3**

    b. Suppose Caitlyn bought the jumbo bag of 80 chocolates. This time, she also shares the chocolates with her teacher. How many chocolates does each person get? **5**

CHAPTER 2 Test

1. Graph {−3, −1, 0, 2, 5} on a number line. **See Teacher's Answer Key.**
2. Name the coordinate of point P. **−2**

Use the coordinate plane shown below for Exercises 3–5. **3. See Teacher's Answer Key.**

3. Copy the grid at the right. Label the x-axis, the y-axis, and the origin.
4. Plot point $A(-2, -1)$ and name the quadrant in which it is located. **III**
5. Plot point $B(4, -3)$ and name the quadrant in which it is located. **IV**

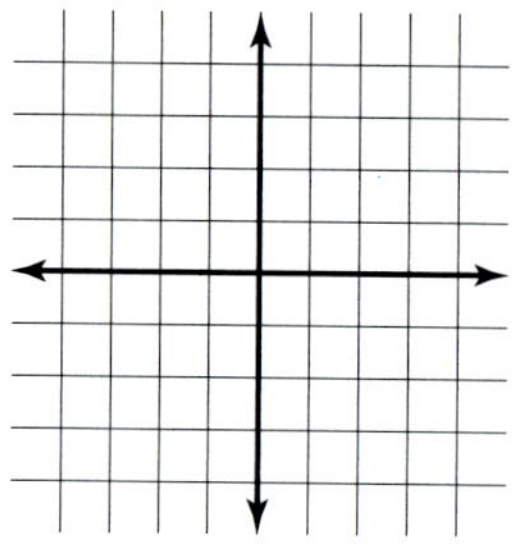

Find each sum or difference.

6. $-4 + 36$ **32**
7. $13 + (-5) + (-6)$ **2**
8. $8 - (-3)$ **11**
9. $4 - 13$ **−9**

10. Find the value of $|y - x|$ if $x = 2$ and $y = -3$. **5**

11. Write −3, 1, −1, 0 in order from least to greatest. **−3, −1, 0, 1**

Simplify.

12. $(-8)(-7)$ **56**
13. $3(-6)$ **−18**
14. $(-9)(4) + (2)(1)$ **−34**
15. $3(-3) - 4(2)$ **−17**
16. $21 \div 7$ **3**
17. $-16 \div 4$ **−4**
18. $\frac{-63}{9}$ **−7**
19. $\frac{-32}{-4}$ **8**

20. **Nutrition** The table shows the amount of sodium a can of regular soda pop contains compared with the amount of sodium a can of diet soda pop contains for three different brands.

    a. Find the differences in sodium for each brand. **25, 5, 15**

    b. Which brand has the greatest difference in sodium? **Brand A**

Amount of Sodium in a Can of Soda Pop (mg)

Brand	Regular	Diet
A	50	25
B	35	30
C	25	10

REVIEW 3-1

Rational Numbers

Rational Number	A rational number is any number that can be expressed as a fraction where the numerator and denominator are integers and the denominator is not zero

Comparison Property	For any two numbers a and b, exactly one of the following sentences is true: $a < b$, $a = b$, or $a > b$

To compare two fractions, you can use **cross products**. Cross products are the products of the diagonal terms.

Comparison Property for Rational Numbers	For any rational numbers $\frac{a}{b}$ and $\frac{c}{d}$, with $b > 0$ and $d > 0$: **1.** if $\frac{a}{b} < \frac{c}{d}$, then $ad < cd$, and **2.** if $ad < bc$, then $\frac{a}{b} < \frac{c}{d}$.

This property also holds if < is replaced by >, ≤, ≥, or =.

Example

Replace ● with <, >, or = to make a true sentence.

$\frac{8}{10}$ ● $\frac{3}{4}$

8(4) ● 10(3)

$32 > 30$

If $32 > 30$, then $\frac{8}{10} > \frac{3}{4}$.

Replace each ● with <, >, or = to make a true sentence.

1. -6 ● 3 **<**

2. $\frac{4}{7}$ ● $\frac{5}{8}$ **<**

3. -3.35 ● -3.40 **>**

Write the numbers in each set from least to greatest.

4. $-\frac{1}{5}$, 0.3, -2.5 **$-2.5, -\frac{1}{5}, 0.3$**

5. $\frac{1}{8}$, -1, $-\frac{9}{7}$ **$-\frac{9}{7}, -1, \frac{1}{8}$**

6. Find a number between $\frac{1}{3}$ and $\frac{7}{9}$. **Sample answer: $\frac{1}{2}$**

7. Which is the better buy: a 15-ounce box of cereal for \$1.99 or a 20-ounce box of cereal for \$2.49? **20-ounce box**

REVIEW 3-2

Adding and Subtracting Rational Numbers

The rules for adding and subtracting integers also apply to adding and subtracting rational numbers.

Rational Number	Form $\frac{a}{b}$
5	$\frac{5}{1}$
$-1\frac{2}{3}$	$-\frac{5}{3}$
0.75	$\frac{3}{4}$

Examples

1 **Find $\left(-3\frac{1}{4}\right) + 6\frac{1}{2}$.**

$$\left(-3\frac{1}{4}\right) + 6\frac{1}{2} = +\left(\left|6\frac{2}{4}\right| - \left|-3\frac{1}{4}\right|\right)$$
$$= +\left(6\frac{2}{4} - 3\frac{1}{4}\right) \text{ or } 3\frac{1}{4}$$

2 **Find $-6.1 - 9.8$.**

$$-6.1 - 9.8 = -6.1 + (-9.8)$$
$$= -15.9$$

3 **Find $-\frac{1}{4} + \frac{5}{6} + \left(-\frac{7}{4}\right)$.**

$$-\frac{1}{4} + \frac{5}{6} + \left(-\frac{7}{4}\right) = \left[-\frac{1}{4} + \left(-\frac{7}{4}\right)\right] + \frac{5}{6}$$ *Comm. & Assoc. Properties of Addition*
$$= -\frac{8}{4} + \frac{5}{6}$$
$$= -\frac{24}{12} + \frac{10}{12}$$ *The LCD is 12.*
$$= -\frac{14}{12} \text{ or } -1\frac{1}{6}$$

Find each sum or difference.

1. $-\frac{4}{13} + \left(-\frac{7}{13}\right)$ $-\frac{11}{13}$
2. $\frac{3}{4} + \left(-\frac{7}{8}\right)$ $-\frac{1}{8}$
3. $-0.01 + 0.06$ **0.05**
4. $\frac{4}{7} - \left(-\frac{2}{7}\right)$ $\frac{6}{7}$
5. $3.97 - 1.55$ **2.42**
6. $-\frac{2}{3} - \frac{4}{9}$ $-\frac{10}{9}$

Evaluate each expression if $a = -0.25$ and $b = \frac{1}{3}$.

7. $a + 0.56$ **0.31**
8. $2 - b$ $\frac{5}{3}$
9. $-3 - a$ **−2.75**
10. $b + \frac{2}{7}$ $\frac{13}{21}$

Find each sum.

11. $-40.1 + 62.7 + (-16.8)$ **5.8**
12. $-13q + (-4q) + 18q$ q
13. $\frac{1}{4} + \left(-\frac{3}{8}\right) + \frac{1}{2}$ $\frac{3}{8}$
14. **Fishing** The saltwater fish record for redeye bass is 8.75 pounds. The record for black sea bass is 0.75 pounds more than the redeye bass record. The record for barred sand bass is 3.75 pounds more than the black sea bass record. What is the record for barred sand bass? **13.25 pounds**

REVIEW **3-3**

Mean, Median, Mode, and Range

In working with statistical data, it is often useful to have one value represent the complete set of data. For example, **measures of central tendency** represent the center or middle of the data. Three measures of central tendency are the **mean, median**, and **mode. Measures of variation** describe the distribution of data. **Range** is a measure of variation.

	Definitions	Numbers
Mean	The mean, or *average*, of a set of data is the sum of the data divided by the number of pieces of data.	Data: 14, 16, 17, 14, 22, 22 $\frac{14 + 16 + 17 + 14 + 22 + 22}{6} = 17.5$
Median	The median of a set of data is the middle number when the data are arranged numerically. For an even number of data items, the median is the mean of the two middle values.	Data: 14, 14, 16, 17, 22, 22 $\frac{16 + 17}{2} = 16.5$
Mode	The mode of a set of data is the number that occurs most often.	Data: 14, 16, 17, 14, 22, 22 There are two modes, 14 and 22.
Range	The range of a set of data is the difference between the greatest and the least values of the set.	Data: 14, 16, 17, 14, 22, 22 $22 - 14$ or 8

Find the mean, median, mode(s), and range for each set of data.

1. 9, 8, 9, 7, 5 **7.6; 8; 9; 4**
2. 2, 1, 0.5, 3, 4.5, 1, 2, 3, 5, 3
3. 10, 18, 18, 18, 10, 10
4. 12, 7, 14, 30 **15.75; 13; none; 23**
5. 5, 7, 7, 13, 11, 5 **8; 7; 5 and 7; 8**
6. 12.4, 6.8, 19.1, 30.4, 7.3

7.

Stem	Leaf
5	3 8 8 9
6	1 2 2 2 4 6
7	7 7 7 8 9 9

6 | 2 = 62

8.

Stem	Leaf
11	1 3 4 5
12	2 2 4 4
13	1 3 3 3 4 5

13 | 3 = 133

9. **Football** The table at the right shows the teams that have played in the Super Bowl most often.
   a. Find the mean, median, mode(s), and range of the data.
   b. Would you choose any measure of central tendency over another to represent these data? Why?

Team	Times in Super Bowl
Dallas Cowboys	8
Denver Broncos	6
Miami Dolphins	5
Pittsburgh Steelers	5
San Francisco 49ers	5
Washington Redskins	5
Buffalo Bills	4
Minnesota Vikings	4
Green Bay Packers	4

Source: *The World Almanac*

2. 2.5; 2.5; 3; 4
3. 14; 14; 10 and 18; 8
6. 15.2; 12.4; none; 23.6
7. 67; 63; 62 and 77; 26
8. 124.6; 124; 133; 24
9a. 5.1; 5; 5; 4
9b. Sample answer: No, all three measures of central tendency represent the data well since they are all either 5 or very close to 5.

REVIEW 3-4

Equations

Mathematical statements with one or more variables are called **open sentences**. Open sentences are **solved** by finding a replacement for the variable that results in a true sentence. The replacement is called a **solution**.

Example 1 Replace y in $5y - 9 = 21$ with the value 6.

$$5y - 9 = 21$$
$$5(6) - 9 = 21$$
$$30 - 9 = 21$$
$$21 = 21 \quad \text{true}$$

Since $y = 6$ makes the sentence $5y - 9 = 21$ true, 6 is a solution.

A set of numbers from which replacements for a variable may be chosen is called a **replacement set**. The set of all replacements for the variable in an open sentence that results in a true sentence is called the **solution set** for the sentence.

A sentence that contains an equals sign, =, is called an **equation** and sometimes may be solved by applying the order of operations.

Example 2 Solve $\frac{3(4 + 5)}{2 \cdot 4 - 3} = w$.

$$\frac{3(4 + 5)}{2 \cdot 4 - 3} = w$$
$$\frac{3(9)}{8 - 3} = w \qquad 4 + 5 = 9;\ 2 \cdot 4 = 8$$
$$\frac{27}{5} = w \qquad 3 \cdot 9 = 27;\ 8 - 3 = 5$$

Find the solution of each equation if the replacement sets are $x = \{-4, 0, 2, 3\}$ and $y = \{-5, -2, 1, 4, 8\}$.

1. $y - 6 = 2$ **8**
2. $6y = -30$ **−5**
3. $1 - x = -1$ **2**
4. $2x + (-5) = -5$ **0**
5. $2x - 3x = 4$ **−4**
6. $\frac{4y + 5}{y} = 9$ **1**

Solve each equation.

7. $d = 2(4 \cdot 5 - 3)$ **34**
8. $m = 5\frac{2}{5} - 1\frac{3}{10}$ **$4\frac{1}{10}$**
9. $\frac{40 - 7}{5 + 6} = z$ **3**

10. **Business** Saundra must sell 4 T-shirts to make a $10 profit.
    a. Write an equation that represents the number of T-shirts she must sell to make a $60 profit.
    b. How many T-shirts would she have to sell? **24**

10a. $t = 60\left(\frac{4}{10}\right)$

REVIEW 3-5 Solving Equations by Using Models

You can use algebra tiles to solve equations.

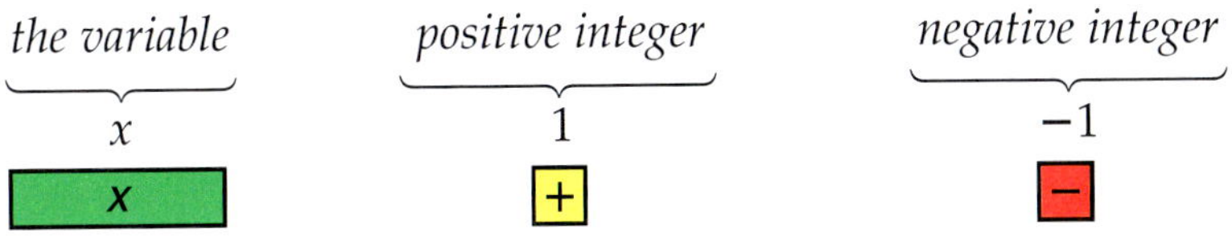

Example

Use algebra tiles to solve $x + 3 = -2$.

Step 1 Model $x + 3 = -2$ by placing 1 green tile and 3 yellow tiles on one side of the mat to represent $x + 3$. Place 2 red tiles on the other side to represent -2.

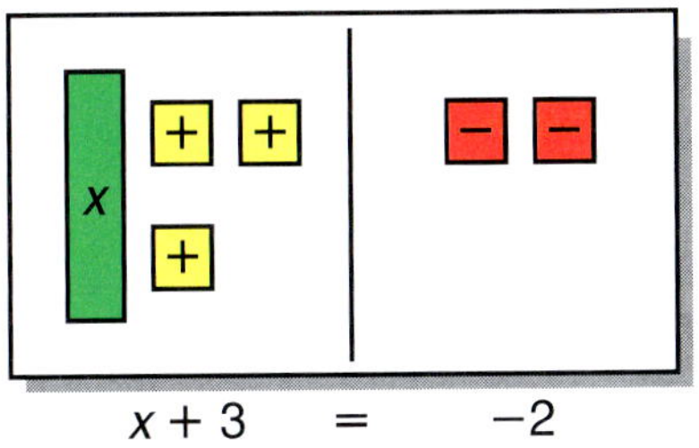

Steps 2 & 3 Since you cannot simply remove 3 yellow tiles from each side of the mat to solve for x, you must add 3 red tiles to each side to make 3 zero pairs on the left side of the mat. Remove the zero pairs.

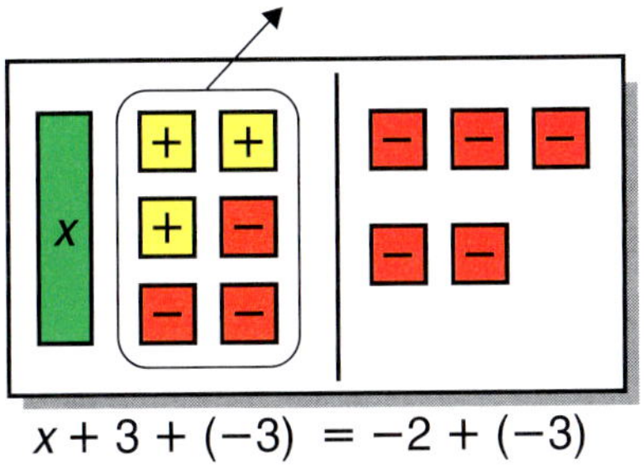

Step 4 The green tile is matched with 5 red tiles. Therefore, $x = -5$.

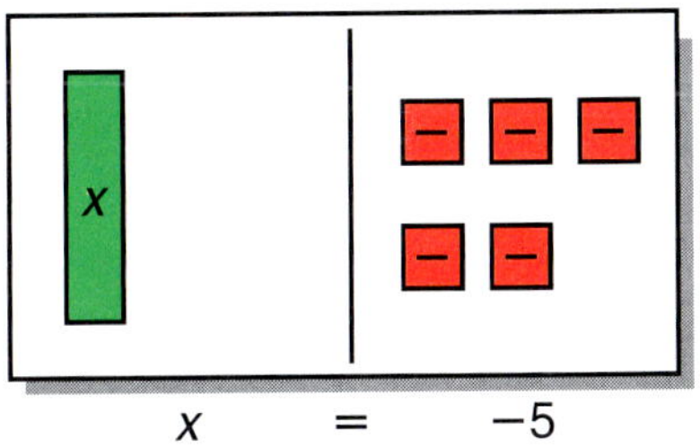

Solve each equation. Use algebra tiles if necessary.

1. $x + (-3) = 1$ **4**
2. $x - 2 = 5$ **7**
3. $x - 4 = 4$ **8**
4. $6 + x = 3$ **−3**
5. $-2 + x = -1$ **1**
6. $x - 1 = -5$ **−4**
7. $x + 7 = 10$ **3**
8. $4 + x = -2$ **−6**
9. $-3 + x = 0$ **3**

10. **Golf** In the 2000 U.S. Open, runners-up Ernie Els and Miguel Angel Jimenez scored 3 above par. This number is 15 more than what the winner of the tournament, Tiger Woods, scored to set a record low for the tournament. **a. Sample answer: $3 = 15 + x$**
    a. Write an equation that represents this situation.
    b. What was Tiger Woods' score? **−12**

REVIEW 3-6 Solving Addition and Subtraction Equations

You can use the Addition and Subtraction Properties of Equality to solve equations. To check, substitute the solution for the variable in the original equation. If the resulting sentence is true, your solution is correct.

Addition Property of Equality	For any numbers a, b, and c, if $a = b$, then $a + c = b + c$.
Subtraction Property of Equality	For any numbers a, b, and c, if $a = b$, then $a - c = b - c$.

Examples

1 **Solve $z - 15 = -3$.**

$$z - 15 = -3$$
$$z - 15 + 15 = -3 + 15$$
$$z = 12$$

Check: $z - 15 = -3$
$12 - 15 \stackrel{?}{=} -3$
$-3 = -3$ ✓

2 **Solve $q + 5 = -2$.**

$$q + 5 = -2$$
$$q + 5 - 5 = -2 - 5$$
$$q = -7$$

Check: $q + 5 = -2$
$-7 + 5 \stackrel{?}{=} -2$
$-2 = -2$ ✓

Sometimes an equation can be solved more easily if it is rewritten first. Recall that subtracting a number is the same as adding its inverse. For example, the equation $b - (-3) = 11$ may be rewritten as $b + 3 = 11$.

Solve each equation. Check your solution.

1. $w - 7 = -3$ **4**
2. $y + 8 = 3$ **−5**
3. $q + 5 = -20$ **−25**
4. $-6 = m + 3$ **−9**
5. $d + (-12) = 4$ **16**
6. $t - (-2) = 14$ **12**
7. $16 + a = -11$ **−27**
8. $62 = 41 + s$ **21**
9. $-7.4 = c + (-1.2)$ **9. −6.2**
10. $-\frac{3}{8} + x = 2\frac{5}{8}$ **3**
11. $-\frac{1}{6} + b = \frac{2}{3}$ $\frac{5}{6}$
12. $\frac{4}{9} = n + \frac{1}{3}$ $\frac{1}{9}$
13. **Clothing Design** A clothing designer is designing a blazer. The blazer will have three buttons on the front. The length of each buttonhole is to be $\frac{1}{8}$-inch longer than the sum of the diameter and the thickness of the button.
    a. Translate the description of the buttonhole length into a formula. **13a.** $\ell = \frac{1}{8} + (d + t)$
    b. Suppose the clothing designer decides that each buttonhole is to be $\frac{15}{16}$-inch long and each button is to be $\frac{5}{8}$-inch in diameter. What thickness should the button be? $\frac{3}{16}$ **inch**

REVIEW **3-7**

Solving Equations Involving Absolute Value

Consider the graph of $|x| = 3$. The distance from 0 to x is 3 units. Thus, $x = -3$ or $x = 3$.

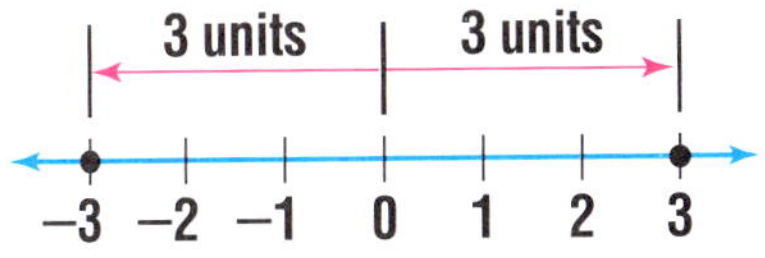

Equations that involve absolute value can be solved by writing them as a compound sentence and solving. Two cases must be considered: when the number inside the absolute value symbol is negative and when it is positive. You can check your solutions by substituting them for the variable in the original equation.

Example

Solve $|r + 9| + 2 = 3$.

First, simplify the expression.

$$|r + 9| + 2 = 3$$
$$|r + 9| + 2 - 2 = 3 - 2 \qquad \text{Subtract 2 from each side.}$$
$$|r + 9| = 1$$

Then write a compound sentence and solve it.

$r + 9 = 1$ or $r + 9 = -1$

$r + 9 - 9 = 1 - 9$ *Subtract 9 from each side.* $r + 9 - 9 = -1 - 9$

$r = -8$ $\qquad$ $r = -10$

The solution set is $\{-10, -8\}$.

Remember that the absolute value of any integer is always positive or zero. So, for example, $|a| = -8$ is never true. Its solution set $\varnothing$ is called the **empty set**.

Solve each equation. Check your solution.

1. $|c| = 13$ $\{\pm 13\}$ 2. $|f| = 100$ $\{\pm 100\}$ 3. $|g| = 0$ $\{0\}$

6. $\{-3, 13\}$ 4. $|x + 7| = 9$ $\{-16, 2\}$ 5. $|-3 + y| = -6$ $\varnothing$ 6. $|b - 5| = 8$

8. $\{-8, 16\}$ 7. $|h| - 10 = -8$ $\{\pm 2\}$ 8. $|-4 + u| = 12$ 9. $|k| + 1 = 0$ $\varnothing$

10. $|7 + p| + 4 = 4$ $\{-7\}$ 11. $-6 + |v - 3| = -2$ $\{-1, 7\}$

12. What is the absolute value of the sum of 9 and -5? **4**
13. The absolute value of 2 and x is 15. What is x? $x = 13$ or $x = -17$

14. **Games** On a game show, contestants guess the price of a prize. If they come within \$100 of the actual price, they win two prizes. If the actual prize is \$24,560.25, write and solve an equation to find the greatest and least amounts a person could bid to win two prizes. $|p - 24,560.25| = 100$; 24,460.25 and 24,660.25

CHAPTER 3 Test

Replace each ● with <, >, or = to make a true sentence.

1. $-6 \bullet -8$ **>**
2. $-3.65 \bullet -3.60$ **<**
3. $\frac{4}{9} \bullet \frac{4}{7}$ **<**
4. Write $\frac{1}{4}$, $\frac{2}{5}$, $\frac{3}{4}$, and $\frac{3}{8}$ in order from least to greatest. **$\frac{1}{4}, \frac{3}{8}, \frac{2}{5}, \frac{3}{4}$**

Find each sum or difference.

5. $2.5 + (-0.3)$ **2.2**
6. $-8.2 - 6.0$ **−14.2**
7. $3.4 - (-0.64)$ **4.04**
8. $-\frac{1}{6} + \left(-\frac{5}{6}\right)$ **−1**
9. $\frac{2}{3} - \left(-\frac{5}{12}\right)$ **$\frac{13}{12}$ or $1\frac{1}{12}$**
10. $\frac{3}{5} + 1\frac{1}{10}$ **$1\frac{7}{10}$**

Solve each equation.

11. $y = (5 - 4) \div 3 \cdot 6$ **2**
12. $q = 3(2.1) - 5.4$ **0.9**
13. $\frac{2 \cdot 1 + 3}{20 \div 2} = r$ **$\frac{1}{2}$**

Solve each equation. Use algebra tiles if necessary.

14. $x + 7 = 2$ **−5**
15. $3 + x = -1$ **4**

Solve each equation. Check your solution.

16. $r - 4 = -48$ **−44**
17. $h + (-9.5) = 1.3$ **10.8**
18. $-\frac{1}{3} + a = \frac{4}{5}$ **$\frac{17}{15}$ or $1\frac{2}{15}$**
19. $|b| + 2 = 8$ **{−6, 6}**
20. $|n - 4| = -10$ **∅**
21. $|s - (-9)| = 5$ **{−14, −4}**

The list below gives the prices of the same pair of sneakers at different stores. Use these data to complete Exercises 22–25.

72, 34, 65, 72, 78, 70, 68, 72, 68

22. Find the mean of the sneaker prices. **$66.56**
23. Find the median of the sneaker prices. **$70**
24. Find the mode of the sneaker prices. **$72**
25. Which measure of central tendency, mean, median, or mode, least represents the sneaker data? Explain. **Mean; only two prices are below the mean, whereas seven are above it.**

REVIEW 4-1

Multiplying Rational Numbers

You can use the same rules for multiplying rational numbers as you did for multiplying integers.

Multiplying Two Rational Numbers	Words	Numbers
Different Signs	The product of two rational numbers having different signs is negative.	$0.2(-5.4) = -1.08$ $-0.2(5.4) = -1.08$
Same Sign	The product of two rational numbers having the same sign is positive.	$8(0.6) = 4.8$ $-8(-0.6) = 4.8$

The rule for multiplying fractions is given below.

Multiplying Fractions	To multiply fractions, multiply the numerators and multiply the denominators.

Example

Find $-\frac{1}{2} \cdot \frac{3}{8}(-4)$.

$$-\frac{1}{2} \cdot \frac{3}{8}(-4) = \left(\frac{-1 \cdot 3}{2 \cdot 8}\right)\left(-\frac{4}{1}\right)$$
$$= -\frac{3}{16}\left(-\frac{4}{1}\right)$$
$$= \frac{3}{4}$$

Find each product.

1. $3 \cdot 2.5$ **7.5**
2. $4.1(6)$ **24.6**
3. $-0.8 \cdot 5$ **-4**
4. $\frac{2}{3} \cdot \frac{1}{4}$ **$\frac{1}{6}$**
5. $\frac{5}{6}(-2)$ **$-\frac{5}{3}$**
6. $\left(-\frac{3}{5}\right)(-55)$ **33**

Simplify each expression.

7. $(-9m)(4.4)$ **$-39.6m$**
8. $-0.2n(\ 0.7)$ **$0.14n$**
9. $-6g \cdot -\frac{3}{2}$ **$9g$**
10. $\left(\frac{2}{7}x\right)\left(\frac{1}{6}y\right)$ **$\frac{1}{21}xy$**

REVIEW 4-2

Counting Outcomes

A result of an experiment or situation is called an **outcome**. You can make a **tree diagram** to find the list of all possible outcomes, commonly called the **sample space**.

Example

A car dealership offers the car options shown in the table. Make a tree diagram to find the sample space.

Option 1	Option 2
radio tape cd	air no air

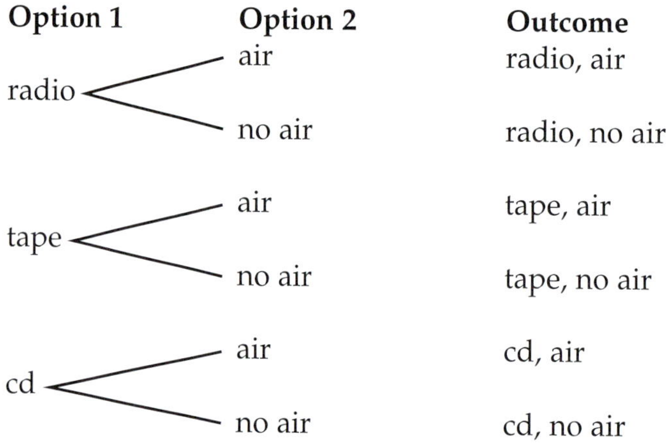

Six possible outcomes make up the sample space.

Find the number of possible outcomes by drawing a tree diagram and listing the sample space. 1–3. See Teacher's Answer Key.

1. two tosses of a coin

2. Three people have been nominated for the positions of secretary or treasurer of the student council. How many different ways can the positions be filled?

Person	Position
Amy Ben Pete	secretary treasurer

3. Four people are swimming on a relay team in which each person swims one of four legs. How many different ways can the team swim the relay?

Person	Leg
Ed Kazu Mario Jay	first second third fourth

REVIEW 4-3

Dividing Rational Numbers

Use the following rules to divide rational numbers.

Rule or Property	Words	Numbers
Dividing Two Rational Numbers	The quotient of two numbers having the same sign is positive. The quotient of two numbers having different signs is negative.	$-08 \div -4 = 0.2$ $-6.3 \div 7 = 0.9$
Multiplicative Inverse Property	The product of a number and its multiplicative inverse is 1.	$\frac{3}{8} \cdot \frac{8}{3} = 1$ $-\frac{2}{5} \cdot -\frac{5}{2} = 1$
Dividing Fractions	To divide a fraction by any nonzero number, multiply by its reciprocal.	$\frac{3}{7} \div \frac{3}{4} = \frac{3}{7} \cdot \frac{4}{3} = \frac{4}{7}$

Since the fraction bar indicates division, you can use the division rules and the Distributive Property to simplify rational expressions.

Example

Find the quotient of $\frac{2}{5}$ and $1\frac{1}{3}$.

$$\frac{2}{5} \div 1\frac{1}{3} = \frac{2}{5} \div \frac{4}{3}$$
$$= \frac{2}{5} \cdot \frac{3}{4}$$
$$= \frac{6}{20} \text{ or } \frac{3}{10}$$

Find each quotient.

1. $5.4 \div 9$ **0.6**
2. $48 \div -1.6$ **−30**
3. $-1.2 \div -6$ **0.2**
4. $-\frac{4}{5} \div \frac{7}{10}$
5. $-\frac{7}{3} \div \frac{1}{2}$
6. $\frac{-96}{-37} \div \frac{96}{37}$ **1**
7. $\frac{2}{3} \div \frac{3}{5}$ **$\frac{10}{9}$ or $1\frac{1}{9}$**
8. $-\frac{1}{5} \div 5$ **$-\frac{1}{25}$**
9. $-\frac{2}{4} \div \frac{3}{6}$ **−1**
10. $\frac{18}{2} \div \frac{1}{3}$ **27**
11. $4 \div \frac{1}{8}$ **32**
12. $\frac{-\frac{3}{7}}{3}$ **$-\frac{1}{7}$**

4. **$-\frac{8}{7}$ or $-1\frac{1}{7}$**
5. **$-\frac{14}{3}$ or $-4\frac{2}{3}$**

13. **Cooking** Theo has a recipe for chocolate chip cookies that makes 6 dozen cookies. He only has enough flour to make 3 dozen cookies.
    a. If the original recipe calls for $2\frac{1}{4}$ cups of margarine, how much margarine will he need to make 3 dozen cookies? **$1\frac{1}{8}$ cups**
    b. If the original recipe calls for $3\frac{1}{2}$ teaspoons of baking powder, how much baking powder will he need to make 3 dozen cookies?
    $1\frac{3}{4}$ cups

REVIEW 4-4

Solving Multiplication and Division Equations

You can solve equations in which a variable has a coefficient by using the Multiplication and Division Properties of Equality.

Division Property of Equality	If you divide each side of an equation by the same nonzero number, the two sides remain equal.
Multiplication Property of Equality	If you multiply each side of an equation by the same number, the two sides remain equal.

Examples

1 **Solve $4x = 28$.**

$4x = 28$

$\frac{4x}{4} = \frac{28}{4}$

$x = 7$

Check: $4x = 28$

$4(7) \stackrel{?}{=} 28$

$28 = 28$ ✓

2 **Solve $\frac{1}{7}p = 3$.**

$\frac{1}{7}p = 3$

$7\left(\frac{1}{7}p\right) = 7(3)$

$p = 21$

Check: $\frac{1}{7}p = 3$

$\frac{1}{7}(21) \stackrel{?}{=} 3$

$3 = 3$ ✓

Solve each equation. Check your solution.

1. $-6z = -72$ **12**
2. $-4q = 48$ **−12**
3. $5r = -35$ **−7**
4. $\frac{1}{7}a = -3$ **−21**
5. $8s = \frac{4}{7}$ $\frac{1}{14}$
6. $\frac{4}{3}t = -\frac{8}{3}$ **−2**

Write an equation and solve.

7. Seven times a number x is 63. What is the number? $7x = 63$; 9
8. One third of a number n is twelve. What is the number? $\frac{1}{3}n = 12$; 36
9. Negative three times a number y is −33. What is the number? $-3y = -33$; 11
10. Tyrone paid \$82.50 for 3 ballet tickets. What is the cost of each ticket t? $3t = 82.50$; \$27.50

Complete.

11. If $5y = 10$, then $7y =$ __?__. **14**
12. If $-3r = 27$, then $4r =$ __?__. **−36**
13. If $8b = -40$, then $-2b =$ __?__. **10**
14. If $3m = 30$, then $6m =$ __?__. **60**

REVIEW 4-5

Solving Multi-Step Equations

To solve some equations you must perform more than one operation. First, determine what operations have been done to the variable. Then undo these operations in the reverse order. In other words, work backward.

Examples

1 **How would you solve $\frac{x}{4} + 5 = 13$?**

$\frac{x}{4} + 5 = 13$ First, x was divided by 4. Then 5 was added.

Step 1 Subtract 5 from each side.

$$\frac{x}{4} + 5 - 5 = 13 - 5$$
$$\frac{x}{4} = 8$$

Step 2 Multiply each side by 4.

$$4\left(\frac{x}{4}\right) = 4(8)$$
$$x = 32$$

2 **$3w + 7 = 34$** *Addition of 7 is indicated.*

$3w + 7 - 7 = 34 - 7$ *Therefore, subtract 7 from each side.*

$3w = 27$ *Multiplication by 3 is also indicated.*

$\frac{3w}{3} = \frac{27}{3}$ *Therefore, divide each side by 3.*

$w = 9$

Check:

$3w + 7 = 34$

$3(9) + 7 \stackrel{?}{=} 34$

$27 + 7 \stackrel{?}{=} 34$

$34 = 34$ ✓

Solve each equation. Check your solution.

1. $5r - 21 = 44$ **13**
2. $6y + 2 = 50$ **8**
3. $9a - 6 = 3$ **1**
4. $\frac{3m}{2} = 9$ **6**
5. $2 = \frac{q + 5}{8}$ **11**
6. $1.4 + 4z = 6.6$ **1.3**
7. $\frac{3}{5}c - 6 = 9$ **25**
8. $\frac{t}{-2} + 7 = -9$ **32**
9. $-6 = \frac{5p - (-2)}{-7}$ **8**

Define a variable, write an equation, and solve each problem. Check your solution.

10. Find three consecutive integers whose sum is 12.
11. Find two consecutive odd integers whose sum is 76.
12. Deborah, Travis, and Tito were each born in one of three consecutive years. The sum of their ages is 45. What are the three ages?

10. $n + (n + 1) + (n + 2) = 12$; 3, 4, 5

11. $n + (n + 2) = 76$; 37, 39

12. $n + (n + 1) + (n + 2) = 45$; 14, 15, and 16 years old

REVIEW

4-6 Solving Equations with Variables on Both Sides

If an equation has variables on both sides, use the Addition or Subtraction Property of Equality to write an equivalent equation that has all the variables on one side. Then solve the equation.

Example

Solve $3z - 4 = -5z - 20$.

$$3z - 4 = -5z - 20$$

$$3z + 5z - 4 = -5z + 5z - 20 \quad \textit{Add 5z to each side.}$$

$$8z - 4 = -20$$

$$8z - 4 + 4 = -20 + 4 \quad \textit{Add 4 to each side.}$$

$$8z = -16$$

$$\frac{8z}{8} = \frac{-16}{8} \quad \textit{Divide each side by 8.}$$

$$z = -2$$

Check:

$$3z - 4 = -5z - 20$$

$$3(-2) - 4 \stackrel{?}{=} 5(-2) - 20$$

$$-6 - 4 \stackrel{?}{=} 10 - 20$$

$$-10 = -10 \checkmark$$

Some equations may have *no solution*, and some equations may have *every number* in their solution set. An equation that is true for every value of the variable is called an **identity**.

Solve each equation. Check your solution.

1. $b + 16 = b - 4$ **no solution**
2. $8 - z = 3z$ **2**
3. $12c - 5c = 4c + 15$ **5**
4. $40 - 5s = 3s$ **5**
5. $2.8w + 5.3 = 3.3w - 0.7$ **12**
6. $m + 9 = 8 - m$ $-\frac{1}{2}$
7. $k + 32 = \frac{1}{2}k - 1$ **−66**
8. $\frac{3}{4}x = 2x + 10$ **−8**
9. $7t + 1 = 5t - 5$ **−3**
10. $3d + 1.1 = 2.3 - d$ **0.3**
11. $-7a + 1 = 1 - 7a$ **identity**
12. $3n - 1 = 9n + 4$ $-\frac{5}{6}$
13. **Sales** Rundell owns a chain of coffee shops. His shops sell both doughnuts and bagels. Over the past few years he has found that the sales of doughnuts have been decreasing by 0.01 million dollars per year, and sales of bagels have been increasing by 0.06 million dollars per year. Last year, Rundell's shops sold 1.6 million dollars in doughnuts and 0.9 million dollars in bagels. If the sales trends continue, after how many years will sales of bagels and doughnuts be equal? **10 years**

REVIEW 4-7

Solving Equations with Grouping Symbols

When an equation contains parentheses or other grouping symbols, first use the Distributive Property to remove the grouping symbols. Then solve the equation as normal.

Example

Solve $4(m + 9) = 3(8 + m)$.

$$4(m + 9) = 3(8 + m)$$

$$4m + 36 = 24 + 3m \quad \textit{Distributive Property}$$

$$4m - 3m + 36 = 24 + 3m - 3m \quad \textit{Subtract 3m from each side.}$$

$$m + 36 = 24$$

$$m + 36 - 36 = 24 - 36 \quad \textit{Subtract 36 from each side.}$$

$$m = -12$$

Check: $4(m + 9) = 3(8 + m)$

$$4(-12 + 9) \stackrel{?}{=} 3(8 + (-12))$$

$$4(-3) \stackrel{?}{=} 3(-4)$$

$$-12 = -12 \quad \checkmark$$

Solve each equation. Check your solution.

1. $5(x - 4) = 20$ **8**
2. $-4 = 2(3t - 8)$ **2**
3. $(h + 1)(-2) = 6$ **−4**
4. $5 = -(b - 9)$ **4**
5. $2z(5 - (-4)) = 18$ **1**
6. $4(2g + 8) = 32$ 0
7. $10 = (y + 7) - 1$ **4**
8. $6(x + 1) - 5 = 13$ **2**
9. $40 - 5s = -2(-1 + 3s)$ **−38**
10. $3(2m - 6) = 2(3m - 9)$ **identity**
11. $k + 11 = \frac{1}{2}(-14k - 10)$ **−2**
12. $\frac{3v + 15}{3} = 6$ **1**
13. The length of a base of the trapezoid shown at the right is twice the length of the other base.

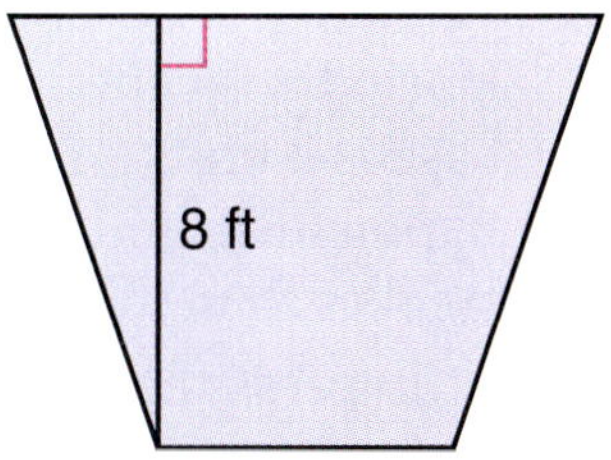

   a. Express the lengths of both bases in terms of x. **x, $2x$**
   b. Find the length of each base of the trapezoid given that the area of the trapezoid is 60 square feet. (*Hint:* The formula for the area of a trapezoid is $A = \frac{1}{2}h(b_1 + b_2)$.) **5 ft, 10 ft**

CHAPTER 4 Test

Find each product or quotient.

1. $-4.7 \cdot -3$ **14.1**
2. $2.1(-0.6)$ **−1.26**
3. $-4.5 \div 9$ **−0.5**
4. $\frac{7}{12} \cdot \frac{3}{4}$ $\frac{7}{16}$
5. $-8 \div -\frac{2}{9}$ **36**
6. $-\frac{1}{2} \div \frac{2}{5}$ $-\frac{5}{4}$

Simplify each expression.

7. $-0.48x(10)$ $-4.8x$
8. $3t\left(\frac{5}{6}\right)$ $\frac{5}{2}t$
9. $\frac{-s}{\frac{5}{3}}$ $-\frac{3s}{5}$

10. Draw a tree diagram to show the different styles of glasses that are possible from the choices given in the table. How many outcomes are in the sample space? **See Teacher's Answer Key.**

Frame	Lens Shape
wire plastic	oval circular rectangular

Solve each equation. Check your solution.

11. $7w = -42$ **−6**
12. $-3z = -3.9$ **1.3**
13. $\frac{1}{8}x = -5$ **−40**
14. $11y + 6 = -38$ **−4**
15. $32 = 7q + 4$ **4**
16. $\frac{3}{7}c + 3 = 9$ **14**
17. $w + 5 = 9w - 3$ **1**
18. $-3(b + 8) = 3(b - 1)$ $-\frac{7}{2}$
19. $4x = -(2x + 48)$ **−8**

20. **Geometry** The side of the triangle labeled h is called the *hypotenuse.* The base is 2 units smaller than the hypotenuse and the height is 4 units smaller than the hypotenuse.

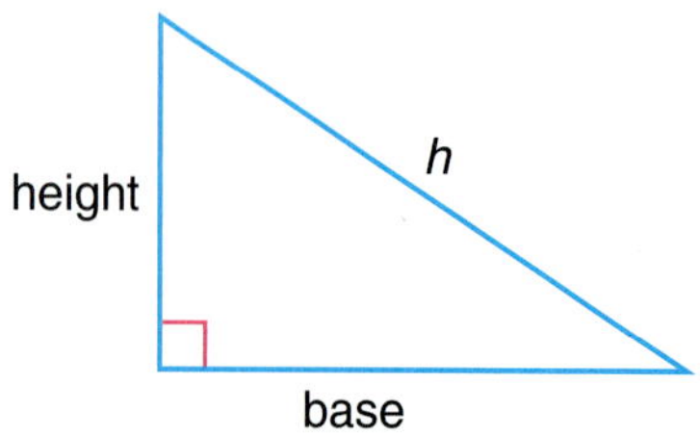

   a. Express the lengths of the base and height of the triangle in terms of the hypotenuse h. **base = h − 2; height = h − 4**

   b. If the perimeter is 24 inches, find the dimensions of the triangle. **base = 8 in., height = 6 in., hypotenuse = 10 in.**

REVIEW 5-1

Solving Proportions

In mathematics, a **ratio** compares two numbers by division. A ratio that compares a number x to a number y can be written in the following ways.

$$x \text{ to } y \qquad x{:}y \qquad \frac{x}{y}$$

When a ratio compares two quantities with different units of measure, that ratio is called a **rate.** For example, a 7°F rise in temperature per hour is a rate and can be expressed as $\frac{7 \text{ degrees}}{1 \text{ hour}}$, or 7 degrees per hour.

Proportions are often used to solve problems involving ratios. You can use the cross products of proportions to solve equations that have the form of a proportion.

Proportion	An equation of the form $\frac{a}{b} = \frac{c}{d}$ stating that two ratios are equal.
Property of Proportions	The cross products of a proportion are equal. If $\frac{a}{b} = \frac{c}{d}$, then $ad = bc$.

Example

Solve $\frac{r}{6} = \frac{14}{3}$.

$\frac{r}{6} = \frac{14}{3}$

$3r = 84$ *Find the cross products.*

$\frac{3r}{3} = \frac{84}{3}$ *Divide each side by 3.*

$r = 28$

The solution is 28.

Solve each proportion.

1. $\frac{3}{11} = \frac{9}{z}$ **33**
2. $\frac{a}{7} = \frac{5}{35}$ **1**
3. $\frac{0.25}{4} = \frac{0.5}{y}$ **8**
4. $\frac{w-1}{9} = \frac{7}{9}$ **8**
5. $\frac{4}{x+2} = \frac{16}{28}$ **5**
6. $\frac{9-2t}{7} = \frac{15}{7}$ **−3**
7. $\frac{2b}{3} = \frac{8}{20}$ $\frac{3}{5}$
8. $\frac{m+3}{-5} = \frac{-m}{-4}$ $-\frac{4}{3}$

Use a proportion to solve each problem.

9. To make a model of the Rio Grande River, Angelica used 1 inch of clay for 50 miles of the actual river's length. Her model river was 38 inches long. How long is the Rio Grande? **1900 miles**
10. Jorge finished 28 math problems in one hour. At that rate, how many hours will it take him to complete 70 math problems?
$2\frac{1}{2}$ **hours**

REVIEW 5-2 Scale Drawings and Models

A **scale drawing** or **scale model** is used to represent an object that is too large or too small to be drawn or built at actual size. The **scale** is the ratio of a length on the drawing or model to the corresponding length of the real object.

Example

In a movie about dinosaurs, the dinosaurs were scale models and so was the sport utility vehicle that the T-Rex overturned. The vehicle was made to the scale of 1 inch to 8 inches. The actual vehicle was about 168 inches long. What was the length of the model vehicle?

$$\frac{1 \text{ inch}}{8 \text{ inches}} = \frac{x \text{ inches}}{168 \text{ inches}}$$ *← model length* / *← actual length*

$168(1) = 8x$ *Find the cross products.*

$168 = 8x$

$\frac{168}{8} = \frac{8x}{8}$

$21 = x$

The length of the model vehicle was 21 inches.

On a map, the scale is 1 inch = 40 miles. Find the actual distance for each map distance.

	From	To	Map Distance	
1.	Buffalo, NY	Rochester, NY	2 inches	80 miles
2.	San Diego, CA	Fresno, CA	8.5 inches	340 miles
3.	Dallas, TX	Amarillo, TX	9 inches	360 miles
4.	Columbus, OH	Springfield, OH	1 inch	40 miles
5.	St. Paul, MN	Duluth, MN	$3\frac{3}{4}$ inches	150 miles
6.	Waterloo, IA	Ames, IA	2.5 inches	100 miles

7. **Photos** In a photograph of a building, one story measures one-fourth inch. If the building is 6 inches tall in the photograph, how many stories high is it? 24

8. **Architecture** On a blueprint of a house, one bathroom is 2.5 inches long. When the house is built, the room will be 10 feet long. Find the scale of the blueprint. 1 inch = 4 feet

REVIEW 5-3

The Percent Proportion

Percent is a ratio that compares a number to 100. A percent problem may be easier to solve if a proportion is used.

Percent Proportion	If P is the percentage, B is the base, and r is the percent, the percent proportion is $\frac{P}{B} = \frac{r}{100}$.

Examples

1 **48 is what percent of 60?**

percentage → / *base* → $\frac{48}{60} = \frac{r}{100}$ ← *rate*

$4800 = 60r$

$80 = r$

48 is 80% of 60.

2 **What number is 30% of 180?**

percentage → / *base* → $\frac{n}{180} = \frac{30}{100}$ ← *rate*

$n = \frac{30}{100}(180)$

$n = 54$

54 is 30% of 180.

Use the percent proportion to find each number.

2. $33\frac{1}{3}\%$

1. Sixty is what percent of 80? **75%**
2. Twenty is what percent of 60?
3. What is 40% of 90? **36**
4. Find 36% of 240. **86.4**
5. Fifty-five is 44% of what number? **125**
6. 8.4 is 12% of what number? **70**
7. On Wednesday, Eric's Machine Shop received a shipment of 24 drill bits. Eric had ordered 40 drill bits. What percent of his order arrived on Wednesday? **60%**
8. Mackenzie received a commission of 6% on the sale of a house. If the amount of her commission was $8400, what was the selling price of the house? **$140,000**
9. According to the book *Are You Normal?*, 33% of Americans are afraid of snakes, 8% are afraid of thunder and lightning, 4% each are afraid of crowds and dogs, 2% of cats, 3% of driving a car or leaving the house, and $\frac{1}{3}$ are afraid of flying.
   a. In a group of 1025 people, about how many are afraid of snakes? **a. about 338**
   b. About how many are afraid of crowds? **41**
   c. About how many are afraid of flying? **about 342**
10. In 1995, 351 million pairs of athletic shoes were purchased. Women accounted for 42% of those sales, and men for about 33% of those sales. About how many pairs of athletic shoes were purchased by women and by men? **women, 147,420,00; men, 115,830,000**

REVIEW 5-4

The Percent Equation

The equation $P = RB$ is called the **percent equation**. In this equation, R is the **rate**. The rate is the decimal form of the percent. The percent equation can be used to solve any percent problem.

Examples

1 What is 20% of 30?

$P = RB$

$= 0.2(30)$

$= 6$

6 is 20% of 30.

2 2 is what percent of 16?

$P = RB$

$2 = R(16)$

$R = \frac{2}{16} = 0.125$ or 12.5%

2 is 12.5% of 16.

Use the percent equation to find each number.

1. What number is 50% of 120? 60
2. Find 25% of 48. 12
3. Find 15% of 12. 1.8
4. 8 is what percent of 24? $33\frac{1}{3}\%$
5. 40 is 40% of what number? 100
6. Find 150% of 76. 114
7. 200 is 80% of what number? 250
8. 45 is what percent of 50? 90%
9. What number is 75% of 180? 135
10. Find 0.3% of 500. 1.5
11. **Jobs** Alissa is a busser at a restaurant. She makes 15% of all the tips her waiter earns. If her waiter earns \$85 in tips, how much money will Alissa receive? \$12.75
12. You can use the formula $I = prt$ to find the interest earned I after depositing an amount p into an account with an interest rate r after so many years t. Suppose you open an account with \$1000 that pays an annual interest rate of 4%. After 2 years, how much interest will your account have earned? \$80
13. **Sales** A sale is running where you can take 30% off the regular price of an item. For a particular shirt regularly priced at \$28, there is a tag that says take an additional 10% off the sale price. How much money will you save if you buy the shirt? \$10.36

REVIEW 5-5

Percent of Change

Some percent problems involve finding a percent of increase or decrease.

Percent of Increase	Percent of Decrease
A basketball that cost \$20 last year costs \$22 this year. The price increased by \$2 since last year. *amount of increase* → / *original price* → $\frac{2}{20} = \frac{r}{100}$ $200 = 20r$ $10 = r$ or $r = 10$ The percent of increase is 10%.	A jacket that originally cost \$80 is now on sale for \$60. *amount of increase* → / *original price* → $\frac{20}{80} = \frac{r}{100}$ $2000 = 80r$ $25 = r$ or $r = 25$ The percent of decrease is 25%.

The **sales tax** on a purchase is a percent of the regular price. To find the total price, you must calculate the amount of sales tax and add it to the regular price. **Discount** is the amount by which the regular price of a purchase is reduced. To find the reduced price, calculate the discount and subtract it from the regular price.

Find the final price of each item to the nearest cent. When there is a discount and sales tax, compute the discount first.

1. compact disc: \$12.00 **\$9.60**
   discount: 20%
2. two concert tickets: \$35.00 **\$29.75**
   student discount: 15%
3. airline ticket: \$348 **\$243.60**
   early booking discount: 30%
4. photo calendar: \$12.95 **\$13.73**
   sales tax: 6%
5. class ring: \$110 **\$103.58**
   group discount: 12%
   sales tax: 7%
6. multimedia software: \$49.95
   discount: 25% **\$39.33**
   sales tax: 5%

Solve each problem. Round to the nearest tenth of a percent.

7. **Consumerism** According to the U.S. Bureau of Economic Analysis, Americans spent \$402.5 billion on recreation in 1995. In 1996, spending on recreation in the United States had grown to \$431.1 billion. What was the percent of increase in spending on recreation from 1995 to 1996? **about 7.1%**
8. **Agriculture** According to the U.S. Department of Agriculture, the number of farms in the United States in 1992 was 1,925,000. By 1997, this number had fallen to 1,912,000 farms. Find the percent of decrease in the number of farms from 1992 to 1997. **about 0.7%**

REVIEW 5-6

Probability and Odds

The **probability** of an event is a ratio that tells how likely it is that the event will take place.

Definition of Probability

$P(\text{event}) = \frac{\text{number of favorable outcomes}}{\text{number of possible outcomes}}$

Example 1

Ms. Michalski picks 8 of the 24 students in her class at random for a special project. What is the probability of being picked?

$$P(\text{being picked}) = \frac{\text{number of students picked}}{\text{total number of students}}$$

The probability of being picked is $\frac{8}{24}$ or $\frac{1}{3}$.

The probability of any event has a value from 0 to 1. If the probability of an event is 0, it is impossible for the event to occur. An event that is certain to occur has a probability of 1. This can be expressed as $0 \leq P(\text{event}) \leq 1$.

The odds of an event occurring is the ratio of the number of ways an event can occur (successes) to the number of ways the event cannot occur (failures).

Definition of Odds

$\text{Odds} = \frac{\text{number of favorable outcomes}}{\text{number of unfavorable outcomes}}$

Example 2

Find the odds that a member of Ms. Michalski's class will be picked for the special project.

Number of successes: 8 Number of failures: 16

Odds of being picked = number of successes : number of failures
= 8:16 or 1:2

Solve.

1. There are 3 brown puppies, 4 white puppies, and 5 spotted puppies in a pen. What is the probability of pulling out a white puppy at random? $\frac{1}{3}$
2. It will rain 5 times in March and snow 7 times. The other days it will be sunny. What is the probability of sun? What are the odds of sun? $\frac{19}{31}$; 19:12

There are 300 freshmen, 250 sophomores, 200 juniors, and 250 seniors at a high school. Solve each problem.

3. If one student is chosen at random, what is the probability that a freshman will be chosen? $\frac{3}{10}$
4. What would be the probability of choosing a freshman if all of the seniors were eliminated? $\frac{2}{5}$
5. What are the odds of choosing a senior at random? 1:3
6. What are the odds that a junior will not be chosen? 4:1

REVIEW 5-7

Compound Events

A **compound event** consists of two or more simple events connected by the words *and* or *or*. Compound events connected by *and* are **independent events.** The outcome of one event does not effect the outcome of the other event.

Probability of Independent Events	$P(A \text{ and } B) = P(A) \cdot P(B)$

Example 1 **Two coins are tossed. Find the probability that they both land heads up.**

$$P(\text{a head and a head}) = P(\text{a head}) \cdot P(\text{a head})$$
$$= \frac{1}{2} \cdot \frac{1}{2} \text{ or } \frac{1}{4}$$

Compound events can also be connected by *or*. In this case, you may have **mutually exclusive** or **inclusive** events. Mutually exclusive events are two events that cannot occur at the same time. Yet, inclusive events can occur at the same time.

Probability of Mutually Exclusive Events	$P(A \text{ or } B) = P(A) + P(B)$
Probability of Inclusive Events	$P(A \text{ or } B) = P(A) + P(B) - P(A \text{ and } B)$

Example 2 **A six-sided die is tossed. What is the probability that the face shows an even number or a 2?**

Since an even number can also be a two, find the probability of two inclusive events.

$P(\text{even number or } 2)$

$$= P(\text{even number}) + P(2) - P(\text{even number and } 2)$$
$$= \frac{3}{6} + \frac{1}{6} - \frac{1}{6} \text{ or } \frac{1}{2}$$

A bag contains 3 white marbles, 3 red marbles, and 2 blue marbles. Find the probability for each situation.

1. P(drawing a white or red marble) $\frac{3}{4}$
2. P(drawing a white or blue marble) $\frac{5}{8}$
3. P(drawing 2 white marbles, with replacement) $\frac{9}{64}$
4. P(drawing 1 marble of each color, with replacement) $\frac{9}{256}$

CHAPTER 5 Test

Solve each proportion.

1. $\frac{4}{8} = \frac{6}{z}$ **12**

2. $\frac{t-1}{7} = \frac{6}{7}$ **7**

3. $\frac{5}{x+2} = \frac{25}{30}$ **4**

Find each number.

4. What number is 20% of 40? **8**

5. Find 3% of 500. **15**

6. 1 is what percent of 20? **5%**

7. 15 is 75% of what number? **20**

8. 150% of 14 is what number? **21**

9. 6 is what percent of 40? **15%**

10. 54.6 is what percent of 130? **42%**

11. What number is 72% of 600? **432**

12. Maps The scale on a map is 1 inch = 15 miles. If the distance between two points on the map is 6 inches, what is the actual distance between the two points? **90 miles**

13. Banking Ms. Morrison learned that $28 was withdrawn from her checking account when she purchased a scarf in Paris for 126 francs. About how much money would have been withdrawn from her account if she spent 172 francs at a restaurant? **$38.22**

14. Surveys In a survey, 29% of respondents reported that they have returned rented videotapes to the store late. If 3200 people responded to the survey, how many returned a videotape late? **928**

15. Shopping The regular cost of a paperback book is $6.95. If all paperbacks are on sale at a 25% discount, what is the sale cost of a paperback book? **$5.21**

16. Auto Maintenance The price of an oil change increased from $19.95 to $22.95. What was the percent of increase? **15%**

17. Odds If the probability in favor of an event is $\frac{8}{15}$, what are the odds of that event occurring? **8:7**

18. Probability A box of sports cards contains 9 baseball cards, 7 football cards, and 4 basketball cards.

a. What is the probability that a baseball card is chosen at random? $\frac{9}{20}$

b. What is the probability of drawing a football or basketball card? $\frac{11}{20}$

c. What is the probability of drawing a football card, replacing it, and then drawing a basketball card? $\frac{7}{100}$

REVIEW
6-1 Relations

A **relation** is a set of ordered pairs. The **domain** is the set of all first coordinates of the ordered pairs, and the **range** is the set of all second coordinates.

Examples

State the domain and range of each relation.

1. $\{(7, 1), (7, 3), (7, 5)\}$ Domain = $\{7\}$; Range = $\{1, 3, 5\}$
2. $\{(2, 4), (3, 6), (4, 4)\}$ Domain = $\{2, 3, 4\}$; Range = $\{4, 6\}$

Relations can be expressed as ordered pairs, tables, and graphs. The relation $\{(-3, 4), (0, 6), (2, -1)\}$ can be expressed in each of the following ways.

Ordered Pairs

$(-3, 4)$
$(0, 6)$
$(2, -1)$

Table

x	y
−3	4
0	6
2	−1

Graph

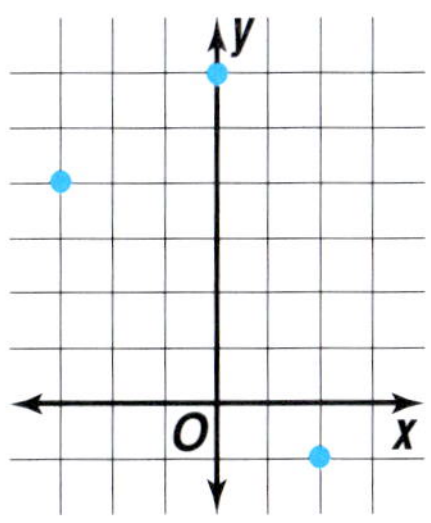

State the domain and range of each relation.

1. $\{(5, 4), (-2, 5), (-3, 0), (4, 5), (5, 0)\}$ D = {−3, −2, 4, 5}; R = {0, 4, 5}
2. $\{(1.25, -0.3), (-14, 12), (6, 1.25)\}$ 2. D = {−14, 1.25, 6}; R = {−0.3, 1.25, 12}
3. $\left\{\left(\frac{1}{3}, \frac{3}{8}\right), \left(-\frac{7}{9}, 4\right), \left(2\frac{1}{5}, -\frac{1}{4}\right)\right\}$ $D = \left\{-\frac{7}{9}, \frac{1}{3}, 2\frac{1}{5}\right\}$; $R = \left\{-\frac{1}{4}, \frac{3}{8}, 4\right\}$

Express each relation as a set of ordered pairs. Then state the domain and the range.

4.

x	y
3	−1
5	4
7	2

{(3, −1), (5, 4), (7, 2)}; D = {3, 5, 7}; R = {−1, 4, 2}

5.

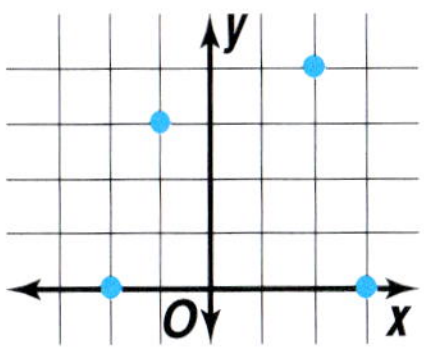

5. {(−2, 0), (−1, 3), (2, 4), (3, 0)}; D = {−2, −1, 2, 3}; R = {0, 3, 4}

6. **Computers** The table gives the percentage of public schools in the United States that have computers capable of using CD-ROM software.

Percentage of Public Schools with CD-ROMs

Year	1995	1996	1997	1998
Percentage	41	54	54	74

a. Determine the domain and range of the relation.
6a. D = {1995, 1996, 1997, 1998}; R = {41, 54, 74}

b. Graph the data. See Teacher's Answer Key.

c. What conclusions might you make from the graph of the data?
The percentage of public schools with CD-ROMs is increasing.

REVIEW 6-2

Equations as Relations

An **equation in two variables** describes a relation since the solutions are ordered pairs.

Solution of an Equation in Two Variables	If a true statement results when the numbers in an ordered pair are substituted into an equation in two variables, then the ordered pair is a solution of an equation.

Example

Solve $y = -x + 4$ if the domain is $\{-4, -2, 6\}$.

Make a table. Substitute each value for x into the equation to determine the corresponding values of y.

x	$-x + 4$	y	(x, y)
−4	−(−4) + 4	8	(−4, 8)
−2	−(−2) + 4	6	(−2, 6)
6	−(6) + 4	−2	(6, −2)

The solution set is $\{(-4, 8), (-2, 6), (6, -2)\}$.

3. $\{(-2, -5), (-1, -4), (0, -3), (1, -2), (2, -1)\}$
4. $\{(-2, 2), (-1, 4), (0, 6), (1, 8), (2, 10)\}$
5. $\{(-2, 11), (-1, 9), (0, 7), (1, 5), (2, 3)\}$
6. $\{(-2, 7), (-1, 4), (0, 1), (1, -2), (2, -5)\}$
7. $\{(-2, -13), (-1, -11), (0, -9), (1, -7), (2, -5)\}$
8. $\left\{\left(-2, 3\frac{1}{2}\right), \left(-1, 3\frac{1}{4}\right), (0, 3), \left(1, 2\frac{3}{4}\right), \left(2, 2\frac{1}{2}\right)\right\}$

Which ordered pairs are solutions of each equation?

1. $a + 3b = 10$ **b, c, d** a. (3, 1) b. (1, 3) c. (10, 0) d. (−2, 4)
2. $5x + 2y = 18$ **a, c** a. (4, −1) b. (3, 1) c. (2, 4) d. (1, 7)

Solve each equation if the domain is $\{-2, -1, 0, 1, 2\}$.

3. $y = x - 3$
4. $y = 2x + 6$
5. $y = -2x + 7$
6. $3x + y = 1$
7. $2x - 9 = y$
8. $y = -\frac{x}{4} + 3$

9. **Cooking** For mashed potatoes, some cooks recommend using one more potato than the number of people to be served.
   a. Write an equation describing the number of potatoes to use when making mashed potatoes. Let x be the number of people to be served. Let y be the number of potatoes to use. $y = x + 1$
   b. Make a graph of the solution set if the domain is $\{3, 4, 5, 6\}$. **See Teacher's Answer Key.**

REVIEW 6-3

Graphing Linear Relations

An equation whose graph is a straight line is called a **linear equation**.

Linear Equation in Standard Form	A **linear equation** is an equation that can be written in the form $Ax + By = C$, where A, B, and C are any real numbers and A and B are not both zero. $Ax + By = C$ is called the standard form if A, B, and C are integers.

Graphing a Linear Equation

1. Solve the equation for one variable.
2. Make a table of values for the variables.
3. Graph the ordered pairs and connect them with a line.

Example

Graph $y + 2x = 2$.

$y + 2x = 2$

$y = 2 - 2x$

x	$2 - 2x$	y	(x, y)
−1	2 − 2(−1)	4	(−1, 4)
0	2 − 2(0)	2	(0, 2)
1	2 − 2(1)	0	(1, 0)

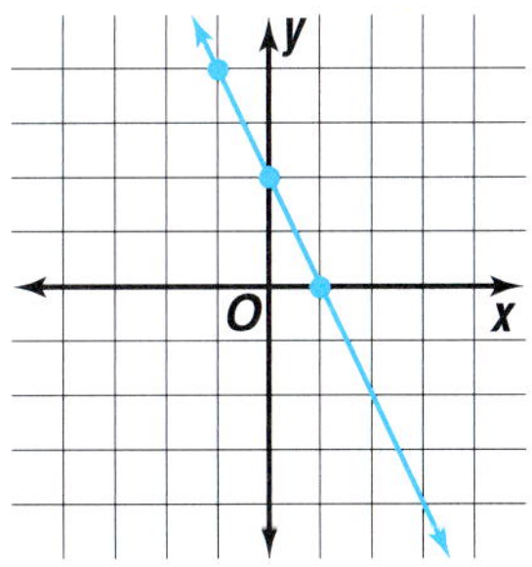

Determine whether each equation is a linear equation. If an equation is linear, identify *A, B,* and *C*.

1. $5x - 3y = 6$
2. $x^2 + 14 = 2y$
3. $9y = 3$

1. yes; $A = 5$, $B = -3$, $C = 6$
2. no
3. yes; $A = 0$, $B = 9$, $C = 3$

Graph each equation. 4–9. See Teacher's Answer Key.

4. $3x + 3y = 9$
5. $x - y = 4$
6. $x + y = -3$
7. $2x - y = 4$
8. $x - y = -2$
9. $\frac{3}{4}x - \frac{1}{4}y = \frac{1}{8}$

10. **Employment** Maria Tomasso works as a sales representative. She receives a salary of \$2200 per month plus a 4% commission on monthly sales. She estimates that her sales in October, November, and December will be \$2000, \$3500, and \$2800.
    a. Graph ordered pairs that represent her incomes y for the three monthly sales figures x. See Teacher's Answer Key.
    b. Will her total monthly income be more than \$2300 in any of the three months? Explain. Yes, for sales of \$2800 and \$3500, her income is greater than \$2300.

REVIEW 6-4 Functions

A special type of relation is called a **function**.

Functions	A function is a relation in which each member of the domain is paired with *exactly* one element of the range.

Example 1 **Is {(4, 2), (7, 1), (5, −1), (3, 2)} a function?**

Since each element of the domain is paired with exactly one element of the range, the relation is a function.

The equation $y = 3x - 1$ can be written in **functional notation** as $f(x) = 3x - 1$. If $x = 4$, then $f(4) = 3(4) - 1$, or 11. Thus, $f(4)$, which is read "f of 4" is a way of referring to the value of y that corresponds to $x = 4$.

Example 2 **If $f(x) = 5x + 4$, find $f(2)$ and $f(-1)$.**

$$\begin{aligned} f(2) &= 5(2) + 4 \\ &= 10 + 4 \\ &= 14 \end{aligned}$$

$$\begin{aligned} f(-1) &= 5(-1) + 4 \\ &= -5 + 4 \\ &= -1 \end{aligned}$$

Determine whether each relation is a function.

1. 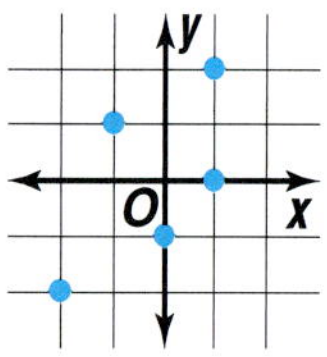 no
2. 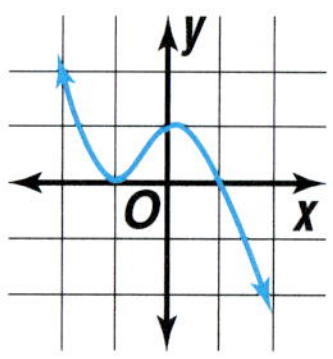 yes
3. 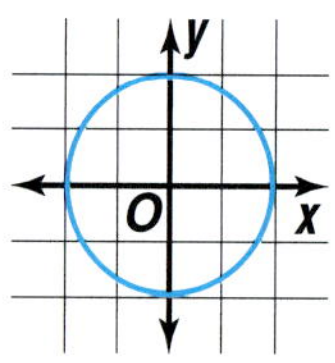 no

4. {(2, 1), (4, 4), (3, 1)}
5. {(7, 3), (6, 4), (7, 5)}
6. {(2, 1), (−2, 1), (3, 1)}
7. {(1, 3), (−1, 0), (3, 6)}
8. {(2, 6), (4, 4), (6, 2)}
9. {(3, 2), (0, 7), (−3, 2)}

4. yes
5. no
6. yes
7. yes
8. yes
9. yes

If $f(x) = 6x + 4$ and $g(x) = 2x - 1$, find each value.

10. $f(2)$ 16
11. $f(-3)$ −14
12. $g(0)$ −1
13. $g(2)$ 3
14. $f(1)$ 10
15. $g(-1)$ −3

16. **Business** Viktor Alessandrovich owns a small ice cream parlor. He has noticed that his daily sales are dependent on the high temperature for the day. The formula for the relationship is $s = 50t + 1000$, where t represents the daily high temperature in degrees Fahrenheit and s is the amount of daily sales in dollars.
   a. Suppose the domain for the function includes the set {60, 70, 80, 90}. Make a table using these values for t and graph the function.
   b. Describe the graph of the function. What trends do you see? As temperature increases, so do daily sales.

16a. See Teacher's Answer Key.

REVIEW 6-5 Direct Variation

If two variables x and y are related by the equation $y = kx$, where k is a nonzero constant, then the equation is called a **direct variation**, and k is called the **constant of variation**. As values for x increase, values for y increase. Similarly, as values for x decrease, values for y decrease.

Example

If y varies directly as x, and $y = 24$ when $x = 6$, find y when $x = 4$.

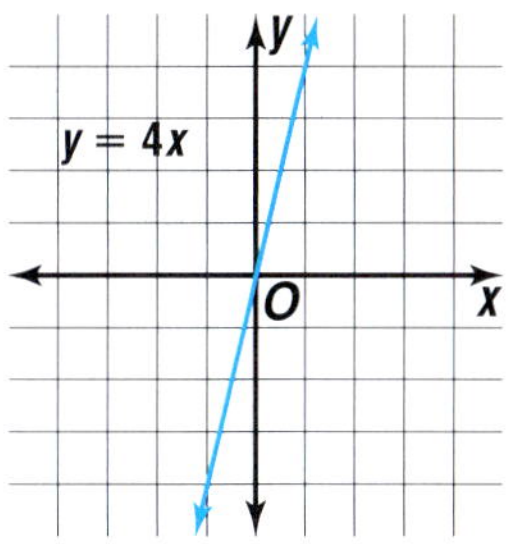

$y = kx$	*Definition of direct variation*
$24 = 6k$	*Replace y with 24 and k with 6.*
$\frac{24}{6} = \frac{6k}{6}$	*Divide each side by 6.*
$4 = k$	

$y = kx$	*Definition of direct variation*
$= 4(4)$	*Replace k with 4 and x with 4.*
$= 16$	

Notice that the graph of direct variation passes through the origin.

Determine whether each equation is a direct variation. Verify the answer with a graph. **1–6. See Teacher's Answer Key for graphs.**

1. $y = x$ **yes**
2. $y = 2x$ **yes**
3. $y = -4x + 1$ **no**
4. $y = -0.25x$ **yes**
5. $y = -1$ **no**
6. $\frac{x}{y} = \frac{1}{3}$ **yes**

Solve. Assume that *y* varies directly as *x*.

7. If $y = 15$ when $x = 3$, find y when $x = 8$. **40**
8. If $y = -49$ when $x = 7$, find x when $y = 91$. **−13**
9. If $y = 0.3$ when $x = 2$, find y when $x = 2.5$. **0.375**
10. If $y = \frac{4}{5}$ when $x = -\frac{3}{4}$, find x when $y = -\frac{7}{10}$. $\frac{21}{32}$

11. **Space** The weight of an object on Mars varies directly as its weight on Earth. An unmanned probe that weighs 500 pounds on Earth weighs 190 pounds on Mars. In the future, NASA hopes to send a manned mission to Mars. How much will an astronaut with gear weighing 220 pounds on Earth weigh on Mars? $83\frac{3}{5}$ **pounds**

REVIEW 6-6

Inverse Variation

If two variables x and y are related by the equation $xy = k$, where k is a nonzero constant, then the equation is called an **inverse variation**, and k is called the **constant of variation**. You may see the equation written as $y = \frac{k}{x}$. As values for x increase, values for y decrease. Similarly, as values for x decrease, values for y increase.

Example

If y varies inversely as x, and $y = 8$ when $x = 9$, find x when $y = 24$.

$xy = k$ — *Definition of inverse variation*

$9(8) = k$ — *Replace x with 9 and y with 8.*

$72 = k$

$xy = k$ — *Definition of inverse variation*

$24x = 72$ — *Replace y with 24 and k with 72.*

$\frac{24x}{24} = \frac{72}{24}$ — *Divide each side by 24.*

$x = 3$

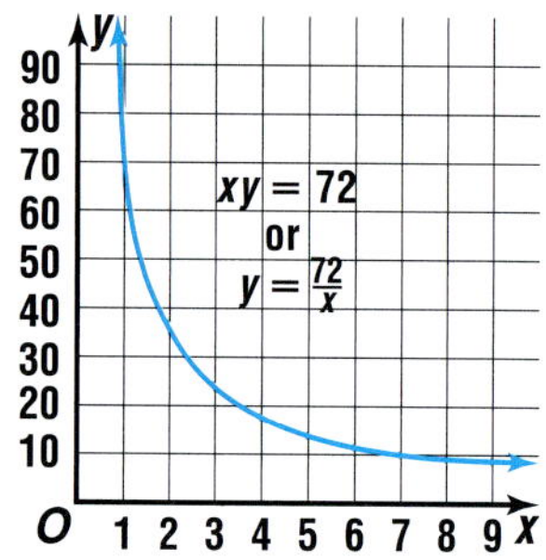

Find the constant of variation. Then write an equation for each statement.

1. y varies inversely as x, and $y = 12$ when $x = 3$. **$36;\ y = \frac{36}{x}$**
2. y varies inversely as x, and $y = -5$ when $x = 7$. **$-35;\ y = -\frac{35}{x}$**
3. y varies inversely as x, and $y = 0.2$ when $x = 10$. **$2;\ y = \frac{2}{x}$**

Solve. Assume that y varies inversely as x.

4. If $y = 10$ when $x = 9$, find y when $x = 3$. **30**
5. If $y = -6$ when $x = 14$, find y when $x = -4$. **21**
6. If $y = 18.1$ when $x = 12.4$, find y when $x = 20$. **11.222**
7. If $y = \frac{3}{8}$ and $x = \frac{1}{9}$, find y when $x = \frac{1}{6}$. **$\frac{1}{4}$**

8. **Electricity** In the formula $V = IR$, where V is voltage from a power source, the current I varies inversely as the resistance R. For a particular light bulb, the resistance is 150 ohms while the current flowing through it is equal to 0.8 amperes. Suppose you would like to dim the light. If you increase the resistance to 200 ohms, find the measure of current. **0.6 ampere**

CHAPTER 6 Test

Use the table to complete Exercises 1 and 2. **1. {(3, 12), (0, 5), (−4, −8)}**

1. Write the relation as a set of ordered pairs.
2. State the domain and range of the relation.
   D = {3, 0, −4}; R = {12, 5, −8}

x	y
3	12
0	5
−4	−8

3. Which ordered pairs are solutions of the equation $2x + 5y = -18$? **c**
   a. $(-3, -2)$ b. $(10, -1)$ c. $(-9, 0)$ d. $(12, -2)$

Solve each equation if the domain is {−2, 0, 2, 4}.

4. $y = x - 6$ **{(−2, −8), (0, −6), (2, −4), (4, −2)}**
5. $y = 3x + 10$ **{(−2, 4), (0, 10), (2, 16), (4, 22)}**
6. $x - 2y = 8$ **{(−2, −5), (0, −4), (2, −3) (4, −2)}**

Graph each equation. **7–9. See Teacher's Answer Key.**

7. $-2x + y = 1$
8. $x + y = -6$
9. $4x - 2y = 0$

Determine whether each relation is a function.

10. 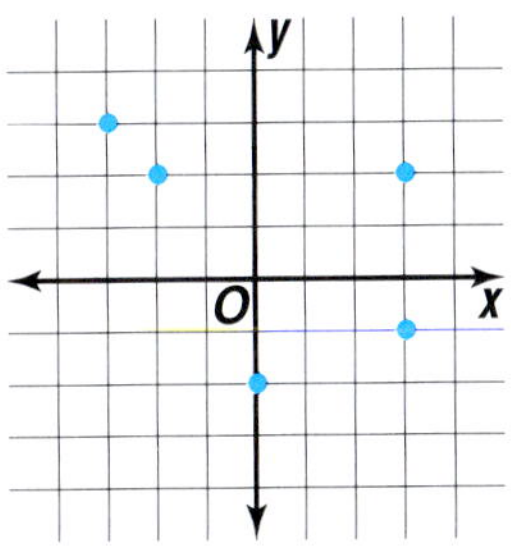

no

11. 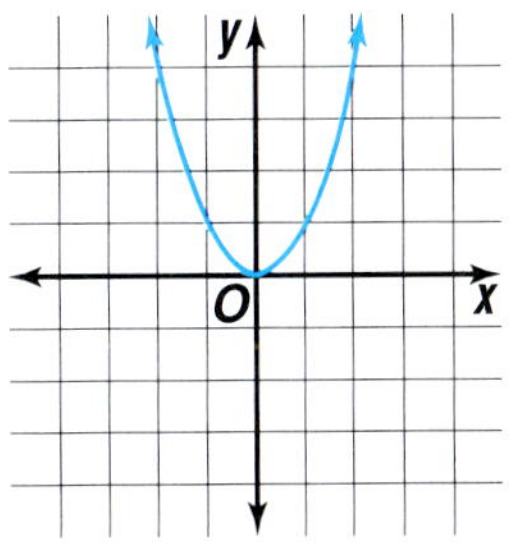

yes

12. $\{(3, 4), (5, 5), (6, 4)\}$ **yes**
13. $\{(1, 0), (2, 0), (5, 0)\}$ **yes**
14. Given $h(x) = -6x$, find $h(-2)$. **12**
15. Given $f(x) = x - 4$, find $f(-5)$. **−9**
16. Given $g(x) = 4x + 5$, find $g(8)$. **37**
17. Given $f(x) = 7x + 12$, find $f(3)$. **33**
18. If y varies directly as x and $y = 12$ when $x = 14$, find y when $x = 49$. **42**
19. If y varies inversely as x and $y = 3$ when $x = 60$, find y when $x = 100$. **1.8**
20. **Variation** The diameter of a circle of light y varies directly as the distance from the light source x. If you stand 2 meters from a wall and shine a flashlight against it, the circle of light on the wall measures 4 meters in diameter.
    a. What does the circle's diameter measure if you stand 3 meters from the wall? four meters? five meters? **6 m, 8 m, 10 m**
    b. Write an equation that represents this direct variation. **$y = 2x$**
    c. Draw a graph of the equation and describe the pattern that you see. **See Teacher's Answer Key.**

REVIEW 7-1 Slope

The ratio of *rise* to *run* is called **slope**. The slope of a line describes its steepness, or rate of change.

On a coordinate plane, a line extending from lower left to upper right has a positive slope. A line extending from upper left to lower right has a negative slope.

The slope of a horizontal line is zero. A vertical line has *no slope.* The slope of a nonvertical line can be determined from the coordinates of any two points on the line.

Slope

The slope of a line is the ratio of the change in the y-coordinates to the corresponding change in the x-coordinates.

$$\text{slope} = \frac{\text{change in } y}{\text{change in } x}$$

Determining Slope Given Two Points

Given the coordinates of two points (x_1, y_1) and (x_2, y_2) on a line, the slope m can be found as follows:

$$m = \frac{y_2 - y_1}{x_2 - x_1}, \text{ where } x_1 \neq x_2.$$

Example

Determine the slope of the line passing through points at (−3, 2) and (6, −4).

$$m = \frac{y_2 - y_1}{x_2 - x_1}$$
$$= \frac{-4 - 2}{6 - (-3)}$$
$$= \frac{-6}{9} = -\frac{2}{3}$$

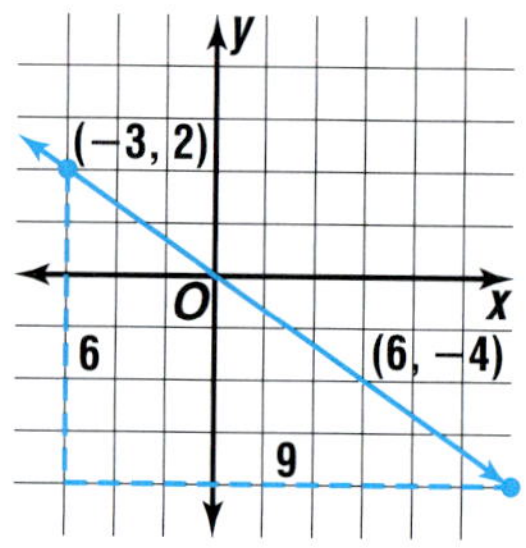

Determine the slope of the line passing through each pair of points whose coordinates are given.

1. $-\frac{1}{3}$
2. $\frac{3}{2}$
3. 1

1.
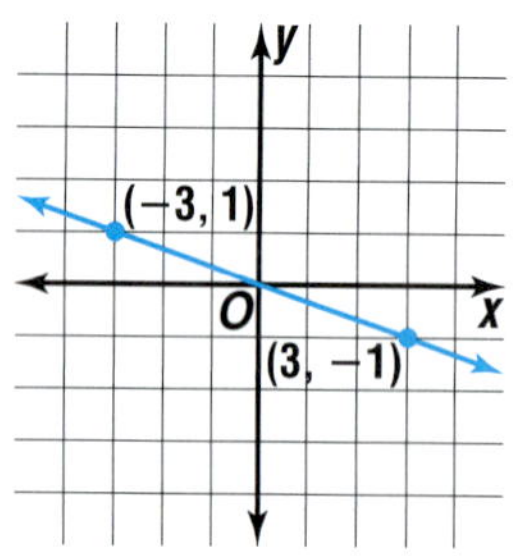

2.
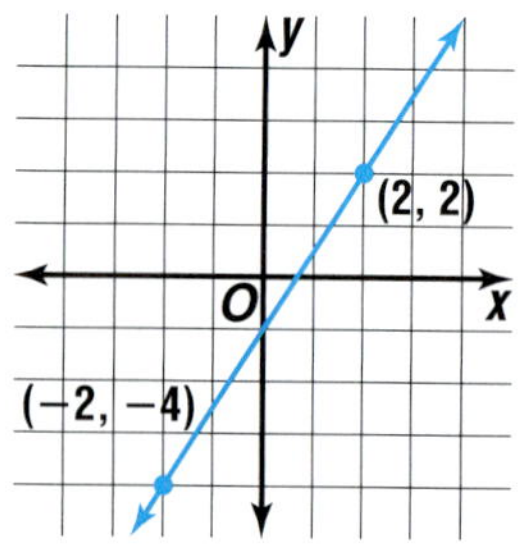

3.
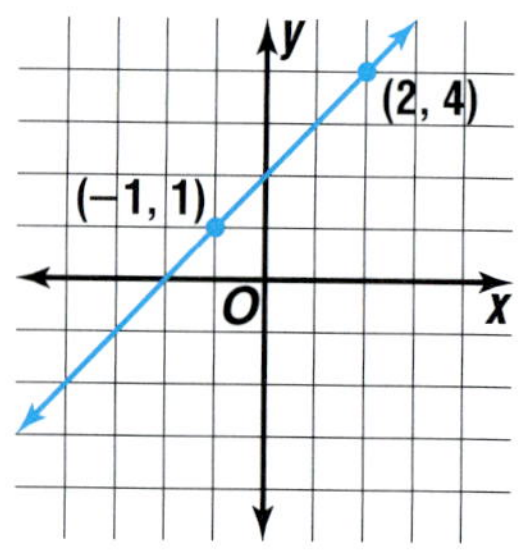

4. (3, 7), (4, 5) -2
5. (10, −2), (6, 3) $-\frac{5}{4}$
6. (3, 0), (9, 1) $\frac{1}{6}$
7. (0, 5), (8, −9) $-\frac{7}{4}$
8. (−2, −2), (−1, 7) 9
9. (4, − 3), (−9, 0) $-\frac{3}{13}$

10. **Road Construction** A portion of the John Scott Memorial Highway in Steubenville, Ohio, has a grade (slope) of 12%. The length of this portion is approximately 1.1 miles. What is the change in elevation from the top of the grade to the bottom of the grade in feet? about 697 ft

REVIEW 7-2

Writing Equations in Point-Slope Form

If you know the slope of a line and the coordinates of one point on the line, you can write an equation of the line by using the **point-slope form**.

Point-Slope Form	Example
For a nonvertical line through the point at (x_1, y_1) with slope m, the point-slope form of a linear equation is $y - y_1 = m(x - x_1)$.	Given: point at $(-1, 3)$ and slope $m = 2$. $(y - 3) = 2(x - (-1))$ $(y - 3) = 2(x + 1)$

Example

Write the point-slope form of an equation of the line that passes through (−3, 8) and has a slope of $-\frac{2}{7}$.

$y - y_1 = m(x - x_1)$

$y - 8 = -\frac{2}{7}(x - (-3))$ or $y - 8 = -\frac{2}{7}(x + 3)$

You can also find an equation of a line if you know the coordinates of two points on the line. First, find the slope of the line. Then write an equation of the line by using the point-slope form.

Write the point-slopze form of an equation for each line passing through the given point and having the given slope.

1. $(5, 6), m = \frac{7}{10}$
2. $(-1, 0), m = 3$
3. $(7, -8), m = -3$
4. $(1, 4), m = -\frac{8}{9}$
5. $(6, -5), m = \frac{1}{6}$
6. $(9, 6), m = 0$

Write the point-slope form of an equation of the line passing through each pair of points.

7. $(7, 5), (-3, 2)$
8. $(-6, 8), (12, 4)$
9. $(4, 4), (2, -8)$
10. $(-9, -4), (4, -2)$
11. $(6, 7), (7, 6)$
12. $(-4, -9), (-6, -3)$

13. **Painting** Vanessa is painting a wall mural. To paint the perspective correctly, she must connect two points on the wall with a line. One point is two feet from the bottom of the painting, three feet from the left side of the painting, and has coordinates (3, 2). The other point is seven feet from the bottom of the painting and ten feet from the left of the painting.
    a. Write the point-slope form of an equation of the line that connects these two points. $y - 2 = \frac{5}{7}(x - 3)$
    b. Vanessa has painted the line on her mural but is worried that the line isn't straight. The line she has painted goes through the point (6, 5). Has she painted the line correctly? Explain. No; the slope of the line through points (3, 2) and (6, 5) is 1 so it does not lie on the line whose slope is $\frac{5}{7}$.

Answers:

1. $y - 6 = \frac{7}{10}(x - 5)$
2. $y = 3(x + 1)$
3. $y + 8 = -3(x - 7)$
4. $y - 4 = -\frac{8}{9}(x - 1)$
5. $y + 5 = \frac{1}{6}(x - 6)$
6. $y = 6$
7. $y - 5 = \frac{3}{10}(x - 7)$
8. $y - 4 = -\frac{2}{9}(x - 12)$
9. $y - 4 = 6(x - 4)$
10. $y + 2 = \frac{2}{13}(x - 4)$
11. $y - 6 = -1(x - 7)$
12. $y + 9 = -3(x + 4)$

REVIEW **7-3**

Writing Equations in Slope-Intercept Form

The x-coordinate of a point where a line crosses the x-axis is called the **x-intercept**. The y-coordinate of the point where the line crosses the y-axis is called the **y-intercept**.

You can always write an equation of a line if you know the slope and y-intercept.

Slope-Intercept Form	Model
Given the slope m and y-intercept b of a line, the slope-intercept form of an equation of the line is $y = mx + b$.	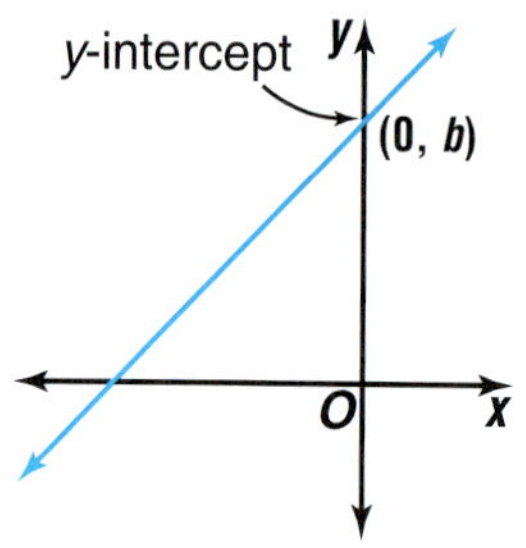

Examples

Write an equation of the line in slope-intercept form for each situation.

1 $m = -2, b = 0$

$y = mx + b$

$y = -2(x) + 0$

$y = -2x$

2 $m = \frac{3}{2}, (-2, 5)$

$y - y_1 = m(x - x_1)$

$y - (5) = \frac{3}{2}(x - (-2))$

$y - 5 = \frac{3}{2}x + 3$

$y = \frac{3}{2}x + 8$

2. $y = 3x + 1$
3. $y = -9x + 8$
4. $y = 2x - 7$
5. $y = \frac{1}{3}x - 7$
6. $y = -x + 6$
7. $y = -\frac{5}{3}x + 10$
8. $y = -5$
9. $y = -2x + 4.5$
11. $y = -\frac{1}{2}x + 3\frac{1}{2}$
12a. $y = \frac{9}{5}x + 32$

Write an equation of the line in slope-intercept form for each situation.

1. $m = 0, b = 5$ $y = 5$
2. $m = 3, b = 1$
3. $m = -9, b = 8$
4. $m = 2, b = -7$
5. $m = \frac{1}{3}, b = -7$
6. $m = -1, (2, 4)$
7. $m = -\frac{5}{3}, (6, 0)$
8. $m = 0, (8, -5)$
9. $m = -2, \left(4, -\frac{7}{2}\right)$
10. $(-3, 0)$ and $(1, 0)$ $y = 0$
11. $(-1, 4)$ and $(3, 2)$

12. **Temperature** Water freezes at 32°F and boils at 212°F. Likewise, water freezes at 0°C and boils at 100°C. Assume that temperature in Fahrenheit y is a linear function of temperature in degrees Celsius x.
    a. Write an equation in slope-intercept form for this function.
    b. What would the temperature be in degrees Fahrenheit if it is 25°C? 77°F

REVIEW 7-4

Scatter Plots

A **scatter plot** is a graph that shows the relationship between paired data. The scatter plot may reveal a pattern, or association, between the paired data.

- Data points that appear to go uphill show a *positive* relationship.
- Data points that appear to go downhill show a *negative* relationship.

The scatter plot at the right represents the relationship between the amount of time Anita spends on her Spanish homework each week and her score on her weekly Spanish quiz. Since the points appear to go uphill, there seems to be a positive relationship between the paired data. In general, the scatter plot seems to show that the more Anita studies, the better her quiz score.

Quiz Score vs. Study Time

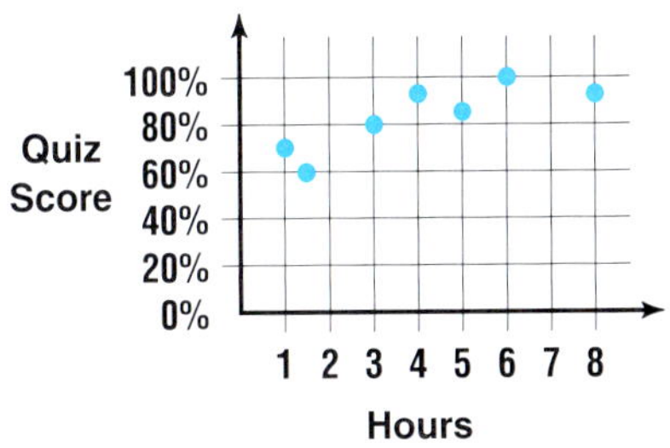

Solve.

1. The table at the right shows the gasoline/mileage record for a certain car. At each gasoline fill-up, the car's owner recorded the amount of gasoline used since the previous fill-up and the distance traveled on that amount of gasoline.

Gasoline/Mileage Record	
Gallons of Gasoline	Miles Traveled
9.8	250
7.9	200
9.2	240
8.0	210
7.5	185
9.3	255
9.5	250

   a. Draw a scatter plot from the data in the table. **See Teacher's Answer Key.**

   b. What are the paired data?

   1b. gallons of gasoline and the miles traveled

   c. Is there a relationship between the gasoline used and the distance driven? **Yes, there is a strong positive correlation.**

2. Alex's biking speed after 10 minutes was 25 mi/h; at 30 minutes, 22 mi/h; at 45 minutes, 20 mi/h; and at 60 minutes, 19 mi/h.

   2a. See Teacher's Answer Key.

   a. Make a scatter plot pairing time biked with biking speed.

   b. How is the data related, positively, negatively, or not at all? **negatively**

3. The table shows the percent of children who are overweight ranked by the hours per day they spend watching television.

Hours of TV Per Day	Percent Overweight
1	12
2	23
3	28
4	30
5 or more	33

   3a. See Teacher's Answer Key.

   a. Draw a scatter plot for these data.

   b. How is the data related, positively, negatively, or not at all? **positively**

   c. What conclusion might you draw from these data? **The more television a child watches, the greater the likelihood that he or she will be overweight.**

REVIEW

7-5 Graphing Linear Equations

There are three methods you can use for graphing equations. You can find two ordered pairs that satisfy the equation, the x- and y-intercepts, or the slope and y-intercept.

Examples

1 **Graph $2x - 3y = 12$ by using the x- and y-intercepts.**

To find the x-intercept, let $y = 0$.

$$2x - 3(0) = 12$$
$$2x = 12$$
$$x = 6$$

To find the y-intercept, let $x = 0$.

$$2(0) - 3y = 12$$
$$-3y = 12$$
$$y = -4$$

Thus, the graph contains the points (6, 0) and (0, −4).

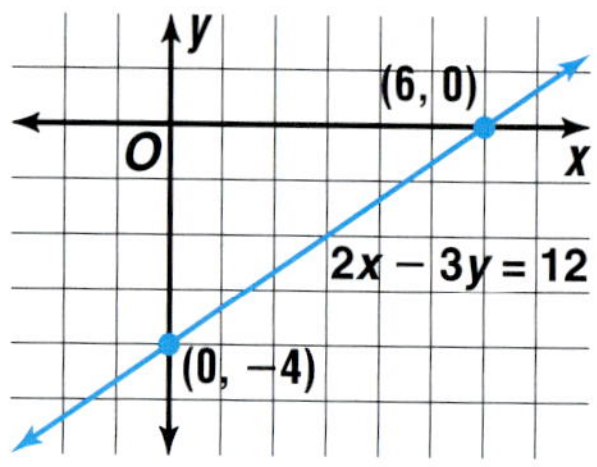

2 **Graph $y = \frac{3}{4}x + 3$ by using the slope and y-intercept.**

The y-intercept is 3, and the slope is $\frac{3}{4}$. Graph a point at (0, 3). From that point, move over 4 units and up 3 units to graph a second point at (4, 6).

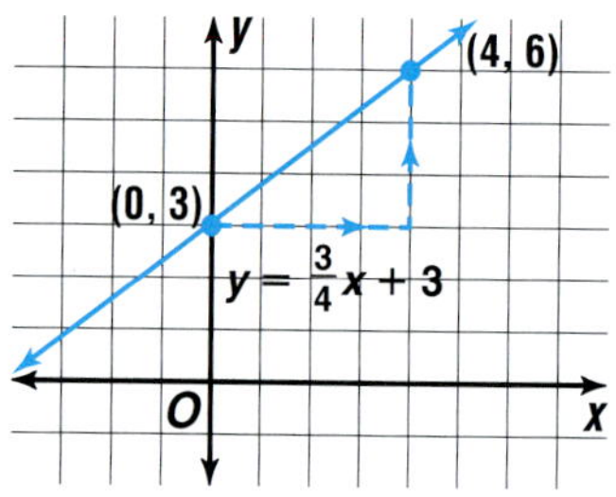

1–4. See Teacher's Answer Key for graphs.

Determine the x-intercept and y-intercept of the graph of each equation. Then graph the equation.

1. $3x - 5y = 15$ **5, −3**
2. $x + 7y = 14$ **14, 2**

Determine the slope and y-intercept of the graph of each equation. Then graph the equation.

3. $y = \frac{1}{5}x + 2$ $m = \frac{1}{5}, b = 2$
4. $y = \frac{3}{4}x - 3$ $m = \frac{3}{4}, b = -3$

5. **Biology** As the temperature increases, the number of times a cricket chirps per minute also increases. The temperature can be approximated by the equation $T = \frac{1}{4}c + 40$, where c represents the number of cricket chirps per minute and T is the temperature in degrees Fahrenheit.
   a. Graph the equation. **See Teacher's Answer Key.**
   b. On a summer evening, you hear a cricket chirping outside your window. Suppose the cricket chirps 132 times per minute. What is the temperature outside? **73°**

REVIEW
7-6 Families of Linear Graphs

Families of graphs are graphs that have at least one characteristic in common. Families of linear graphs are those with the same slope or those with the same x- or y-intercept. These families of graphs are illustrated below.

same slope

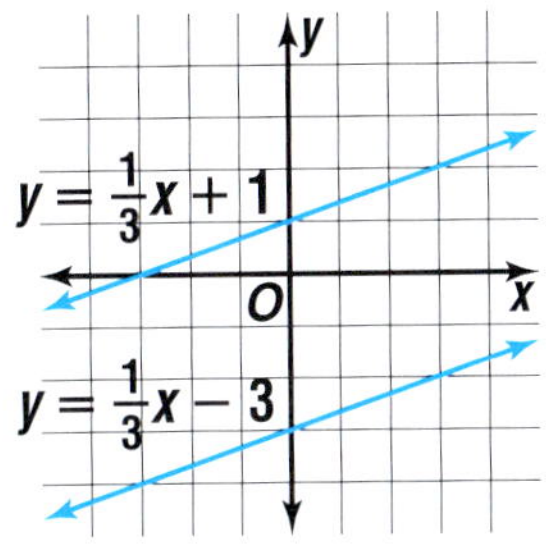

same y-intercept

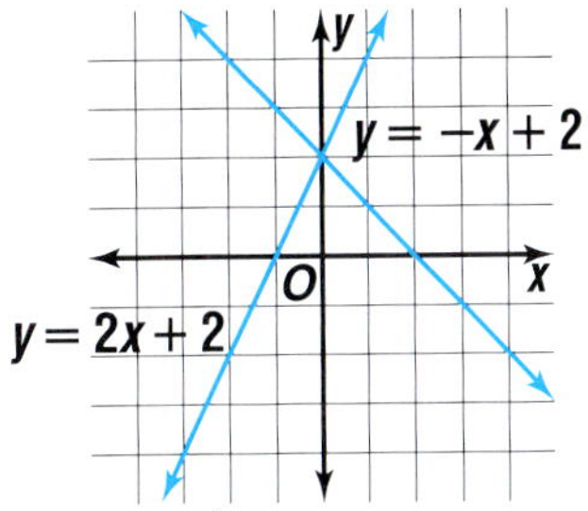

same x-intercept

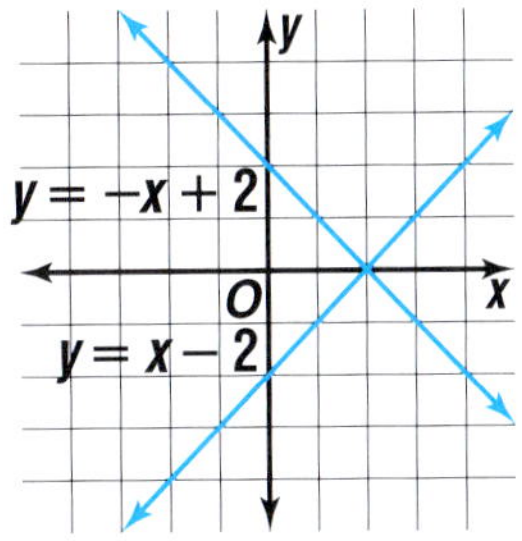

Example

Graph the pair of equations. Describe any similarities or differences. Explain why they are a family of graphs.

$$y = \frac{5}{3}x + 3$$
$$y = -\frac{5}{3}x + 3$$

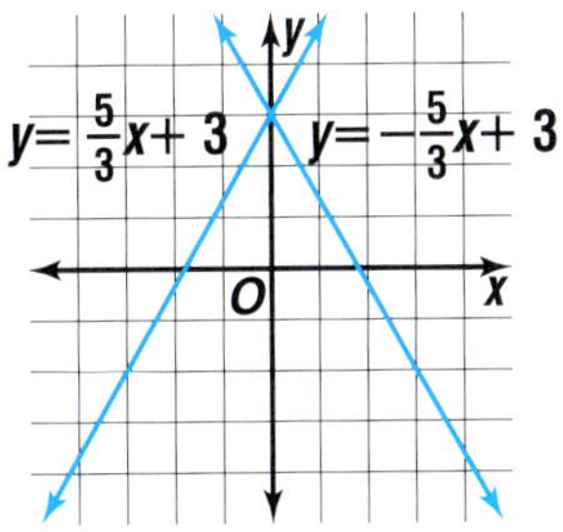

The graphs have slopes equal to $\frac{5}{3}$ and $-\frac{5}{3}$, respectively.

They are a family of graphs because the y-intercept of each line is 3.

1–6. See Teacher's Answer Key.

Graph each pair of equations. Describe any similarities or differences. Explain why they are a family of graphs.

1. $y = 4x - 5$
$y = 4x$

2. $y = 3x + 1$
$y = 3x - 1$

3. $y = 2x - 3$
$y = 3 + 2x$

4. $y = 3x + 2$
$y = x + 2$

5. $y = -\frac{1}{3}x + 7$
$y = -\frac{2}{3}x + 7$

6. $y = \frac{4}{5}x - 4$
$y = -x + 5$

7. Manufacturing Companies often order supplies in bulk from vendors. The graph represents the cost of large sheets of plastic for two different vendors.

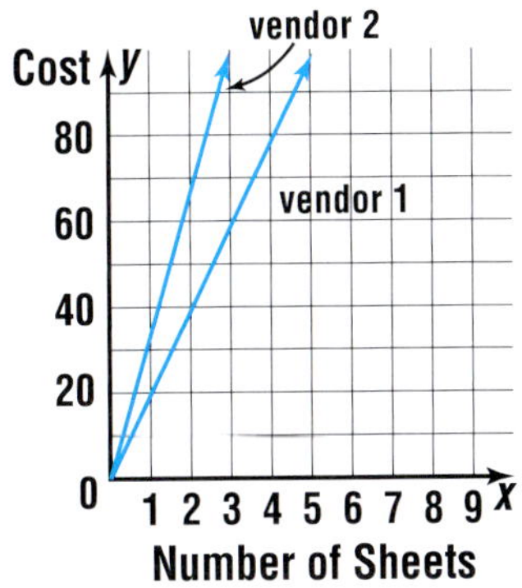

a. Are these graphs a family of graphs? Explain.

b. What does the slope represent?

c. Which vendor's plastic sheets cost more? Explain. vendor 2

7a. Yes, they have the same y-intercept.

7b. cost per plastic sheet

REVIEW 7-7

Parallel and Perpendicular Lines

Two special types of graphs are described below.

Parallel Lines	Perpendicular Lines
If two lines have the same slope, then they are parallel. 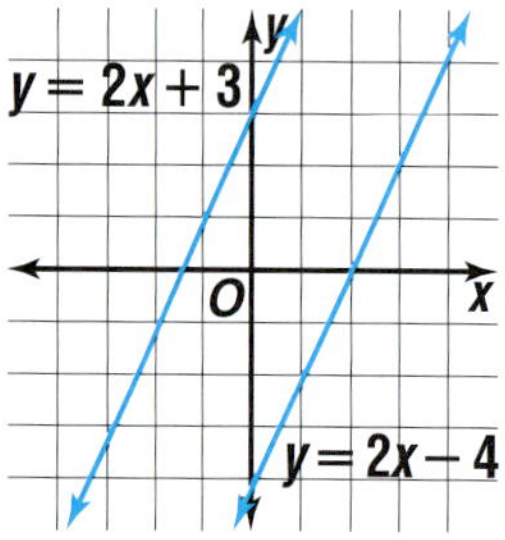	If the products of the slopes of two lines is -1, then the lines are perpendicular. 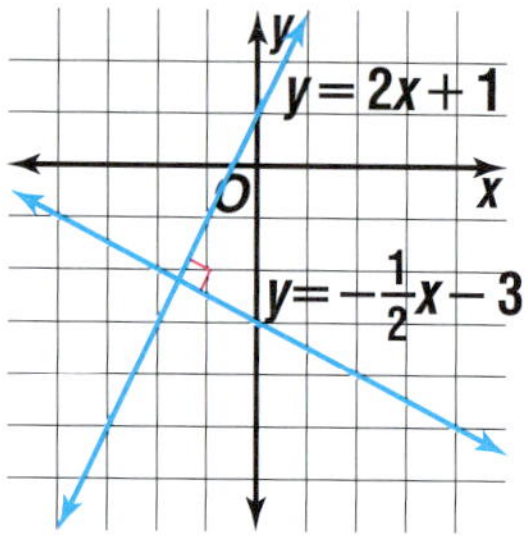

Example

Write an equation in slope-intercept form of the line that is parallel to the graph of $-2y = 6x + 1$ and passes through (3, 0).

Write $-2y = 6x + 1$ in slope-intercept form as $y = -3x - 0.5$.
The slope is -3. The slope of the equation of a parallel line will also be -3 and will go through the point at (3, 0).

$$y - y_1 = m(x - x_1)$$
$$y - (0) = -3(x - 3)$$
$$y = -3x + 9$$

Determine whether the graphs of each pair of equations are *parallel*, *perpendicular*, or *neither*.

1. $y = 2x + 3$, $y = -2x + 3$ **neither**
2. $y = x$, $y = x - 5$ **parallel**
3. $y = 3x - 8$, $3y = -x + 6$

3. perpendicular

Write an equation in slope-intercept form of the line that is parallel to the graph of each equation and passes through the given point. Repeat for a line that is perpendicular to the graph of each equation.

4. $y = 4x - 5$; (1, 6)
5. $y = 5x + 12$; $(-6, 2)$
6. $y = -\frac{1}{3}x + 7$; (8, 0)
7. $x + y = 0$; $(3, -2)$

4. $y = 4x + 2$; $y = -\frac{1}{4}x + 6\frac{1}{4}$
5. $y = 5x + 32$; $y = -\frac{1}{5}x + \frac{4}{5}$
6. $y = -\frac{1}{3}x + \frac{8}{3}$; $y = 3x - 24$
7. $y = -x + 1$; $y = x - 5$

8. **Transportation** In an aerial photograph, one rail in a set of railroad tracks has the equation $y = 0.75x + 3.5$. Find the equation of the other rail. **$y = 0.75x + 8$**

CHAPTER 7 Test

Determine the slope of the line that passes through each pair of points.

1. (6, 9), (1, 10) $-\frac{1}{5}$

2. (8, −3), (2, 5) $-\frac{4}{3}$

3. 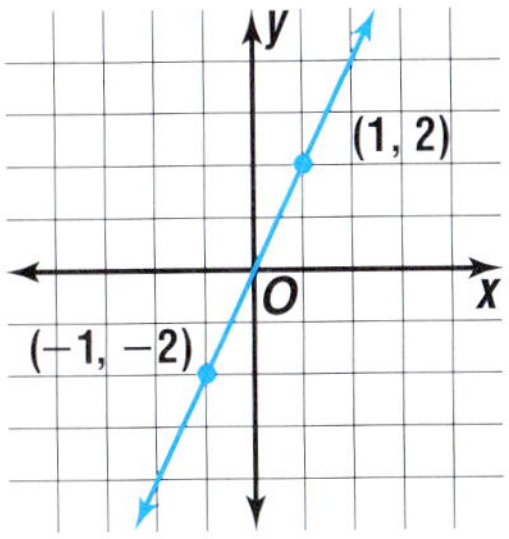

2

4. 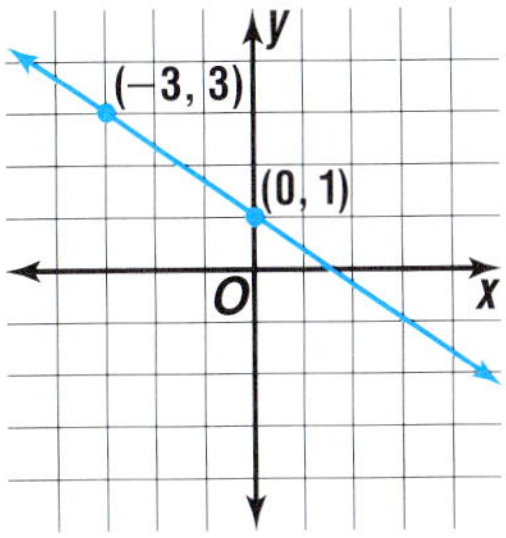

$-\frac{2}{3}$

Write the point-slope form of an equation of the line satisfying the given conditions.

5. passes through (1, 3) and has a slope of $\frac{5}{8}$ $y - 3 = \frac{5}{8}(x - 1)$

6. passes through (12, 3) and has a slope of 0 $y = 3$

7. passes through (4, 7) and (−3, 2) $y - 7 = \frac{5}{7}(x - 4)$ or $y - 2 = \frac{5}{7}(x + 3)$

The value of U.S. exports and imports (in billions of dollars) for selected years from 1970 to 1994 are listed in the table at the right. Use the data for Exercises 8 and 9.

Year	Exports	Imports
1970	43	40
1975	108	99
1980	221	245
1985	213	345
1990	394	495
1995	585	743
1996	625	795
1997	689	871
1998	682	912

Source: U.S. International Trade Administration

8. Make a scatter plot of the data with the value of the exports on the horizontal axis and the value of the imports on the vertical axis. **See Teacher's Answer Key.**

9. Does the data have a *positive*, *negative*, or *no* correlation? **positive**

10. Write an equation in slope-intercept form of a line that has a slope of −4 and *y*-intercept of 7. $y = -4x + 7$

11. Find the slope and *y*-intercept of the graph of $2x + 0.5y = 1$. $m = -4$, $b = 2$

12. Find the *x*- and *y*-intercepts of the graph of $9x - 3y = 5$. $\frac{5}{9}$, $-\frac{5}{3}$

13. Graph $4x - y = 8$ by using the *x*- and *y*-intercepts. **See Teacher's Answer Key.**

14. Graph $y = \frac{3}{5}x - 4$ by using the slope and *y*-intercept. **See Teacher's Answer Key.**

15. Write an equation in slope-intercept form of the line that is parallel to the graph of $y = -\frac{3}{4}x + 6$ and passes through (−8, −5). $y = -\frac{3}{4}x - 11$

16. Write an equation in slope-intercept form of the line that is perpendicular to the graph of $y = 3x + 17$ and passes through (−7, 4). $y = -\frac{1}{3}x + 1\frac{2}{3}$

Posttest

Circle the letter of the correct answer.

1. Write the expression *four less than the quotient of m and n.* **A**

A $\frac{m}{n} - 4$

B $mn - 4$

C $4 - \frac{m}{n}$

D $4 - mn$

2. If $-3 - (-8) = y$, what is the value of y? **H**

F -11

G -5

H 5

I 11

3. Find $-4\frac{3}{8} - 2\frac{3}{4}$. **C**

A $-1\frac{3}{4}$

B $-6\frac{1}{2}$

C $-7\frac{1}{8}$

D $-8\frac{3}{8}$

4. A store's monthly CD sales are shown in the graph. If the average cost of a CD is \$13, including tax, what would be a reasonable sales total for the month of August? **I**

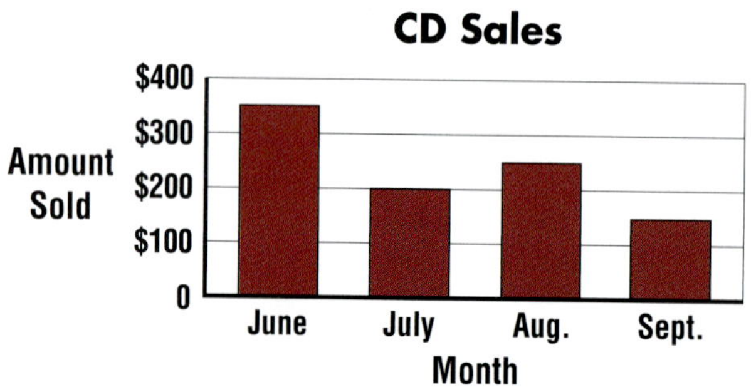

F \$2000

G \$2600

H \$3200

I \$4000

5. Which property is shown by $3(a + 6) = (3 \cdot a) + (3 \cdot 6)$? **D**

A Associative Property of Addition

B Associative Property of Multiplication

C Commutative Property of Addition

D Distributive Property

6. Which ordered pair names point G? **F**

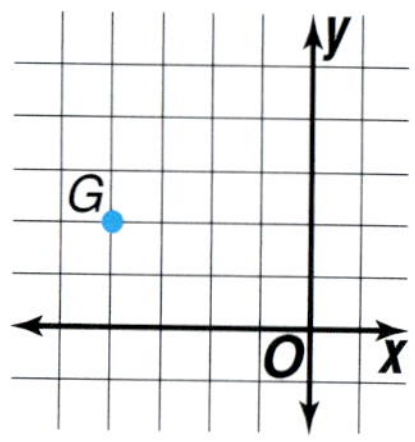

F $(-4, 2)$

G $(4, 2)$

H $(2, -4)$

I $(2, 4)$

7. Evaluate $3[-(12 \div 4)] + 7$. **B**

A -5

B -2

C 2

D 16

8. Simplify $(-5a)(-3b)$. **I**

F $-15ab$

G $-8ab$

H $8ab$

I $15ab$

9. Solve $-2.83 + z = 0.47$. **A**

A 3.3

B 3.2

C 2.3

D -2.36

10. Solve $|x| + 6 = 3$. **I**

F $\{3, -3\}$

G $\{3\}$

H $\{-3\}$

I $\varnothing$

Posttest

11. The table lists the states having the most Alpine ski areas in operation. What is the median of the data? **A**

State	Amount
New York	54
Pennsylvania	34
California	32
Michigan	40
Wisconsin	38

A 38
B 39.6
C 40
D 54

12. Find the value of n in the equation $-28 = \left(-3\frac{1}{2}\right)n$. **I**

F -8
G 2
H 6
I 8

13. Julie and her three friends purchased $2\frac{2}{3}$ pounds of nuts. How much will each person get if they split the nuts equally? **C**

A $\frac{8}{9}$ lb
B $\frac{3}{4}$ lb
C $\frac{2}{3}$ lb
D $\frac{1}{3}$ lb

14. What is the solution of $4(2.5 - n) + 6 = 5n - n$? **G**

F -2
G 2
H 4
I no solution

15. In 10 seconds, light travels about 1.86 million miles. The distance from the sun to Mercury is about 36 million miles. Which proportion could be used to find t, the time it takes for light to travel from the sun to Mercury? **A**

A $\frac{1.86}{10} = \frac{36}{t}$
B $\frac{1.86}{36} = \frac{t}{10}$
C $\frac{1}{t} = \frac{1.86}{36}$
D $\frac{t}{36} = \frac{1.86}{10}$

16. Suppose one disk is chosen at random from the disks shown. What is the probability of selecting an even number greater than 5? **G**

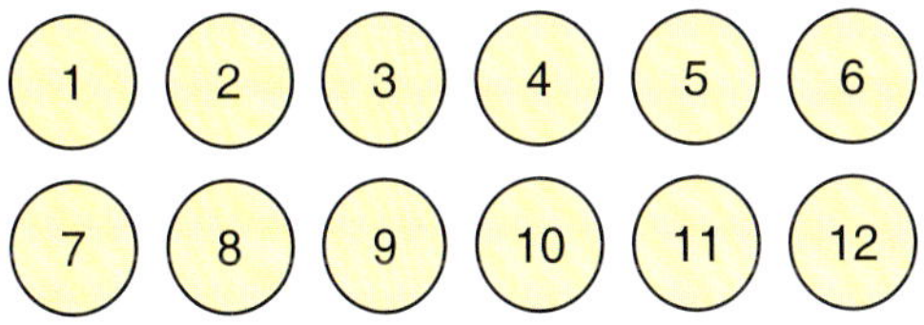

F $\frac{1}{4}$
G $\frac{1}{3}$
H $\frac{1}{2}$
I $\frac{7}{12}$

17. What is the domain of the relation shown in the table? **C**

x	y
-1	5
2	-2
8	1
-3	-1
0	3

A $\{-1, 2, 1, -1, 3\}$
B $\{5, -2, 1, -1, 3\}$
C $\{-1, 2, 8, -3, 0\}$
D $\{-1, 2, 8, -3\}$

Posttest

18. 72 is 18% of what number? G
   - **F** 250
   - **G** 400
   - **H** 1296
   - **I** 4000

19. Between 1987 and 1997, the number of students taking a college-prep course increased from 38% to 61%. Find the percent of increase to the nearest percent. B
   - **A** 58%
   - **B** 61%
   - **C** 62%
   - **D** 165%

20. Which relation is *not* a function? G

   F

x	6	−1	0	6	1
y	4	0	5	4	3

   G

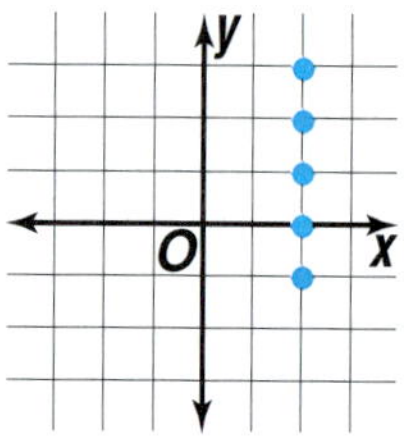

   H $\{(1, 1), (2, 2), (3, 3), (4, 4)\}$

   I

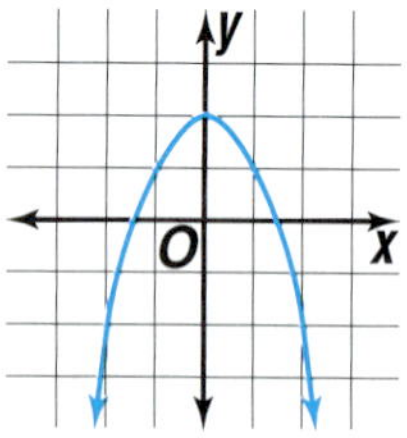

21. Suppose y varies inversely as x. If $y = 16$ when $x = -6$, what is x when $y = 24$? C
   - **A** −14
   - **B** −9
   - **C** −4
   - **D** 4

22. Find the x-intercept for $3x - 2y = 8$. G
   - **F** $(0, 0)$
   - **G** $\left(2\frac{2}{3}, 0\right)$
   - **H** $(0, -4)$
   - **I** $\left(-2\frac{1}{3}, 0\right)$

23. Find the slope of the line that passes through (4, 2) and (−4, 4). B
   - **A** −4
   - **B** $-\frac{1}{4}$
   - **C** $\frac{1}{4}$
   - **D** 4

24. Find the equation of the line through $(-6, 3)$ with slope $\frac{2}{3}$. H
   - **F** $y = \frac{2}{3}x - 8$
   - **G** $y = \frac{2}{3}x + 4$
   - **H** $y = \frac{2}{3}x + 7$
   - **I** $y = \frac{2}{3}x - 1$

25. What is the slope of a line that is perpendicular to the graph of $3x + 7y = 4$? D
   - **A** $-\frac{7}{3}$
   - **B** $\frac{1}{3}$
   - **C** $\frac{3}{7}$
   - **D** $\frac{7}{3}$

Chapter B Teacher's Answer Key

Page B10 Review Lesson 1-7

2a.

Stem	Leaf
2	6
3	9
4	
5	2 5 8
6	4 4
7	2 8 9

5 | 2 = $52

Page B12 Review Lesson 2-1

7.

8.

9.

10.

11.

Page B13 Review Lesson 2-2

1–6.
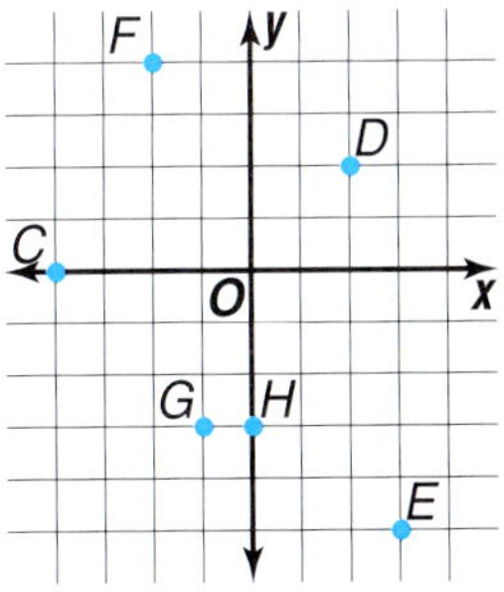

Page B18 Chapter 2 Test

1.

3.
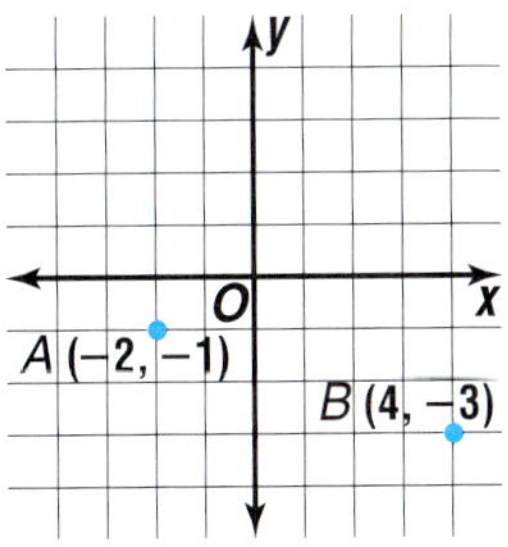

Page B28 Review Lesson 4-2

1.

First Toss	Second Toss	Outcome
head	head	head, head
	tail	head, tail
tail	head	tail, head
	tail	tail, tail

There are 4 outcomes.

2.

Secretary	Treasurer	Outcome
Amy (A)	Ben	A, B
	Pete	A, P
Ben (B)	Amy	B, A
	Pete	B, P
Pete (P)	Amy	P, A
	Ben	P, B

There are 6 outcomes.

3.

Leg 1	Leg 2	Leg 3	Leg 4	Outcome
Ed	Kazu	Mario	Jay	E, K, M, J
		Jay	Mario	E, K, J, M
	Mario	Kazu	Jay	E, M, K, J
		Jay	Kazu	E, M, J, K
	Jay	Kazu	Mario	E, J, K, M
		Mario	Kazu	E, J, M, K
Kazu	Ed	Mario	Jay	K, E, M, J
		Jay	Mario	K, E, J, M
	Mario	Ed	Jay	K, M, E, J
		Jay	Ed	K, M, J, E
	Jay	Ed	Mario	K, J, E, M
		Mario	Ed	K, J, M, E
Mario	Ed	Kazu	Jay	M, E, K, J
		Jay	Kazu	M, E, J, K
	Kazu	Ed	Jay	M, K, E, J
		Jay	Ed	M, K, J, E
	Jay	Ed	Kazu	M, J, E, K
		Kazu	Ed	M, J, K, E

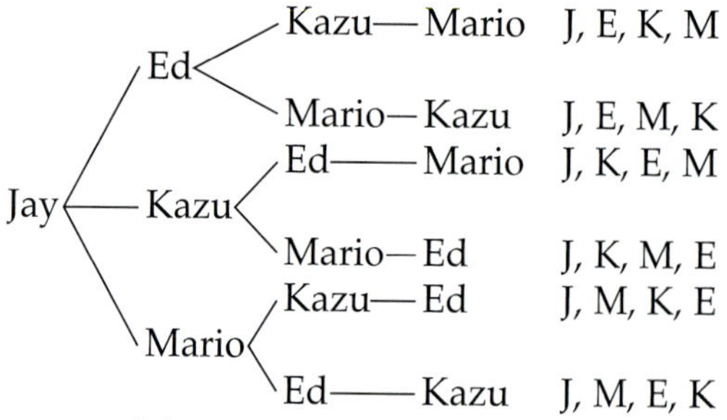

There are 24 outcomes.

Page B34 Chapter 4 Test

10.

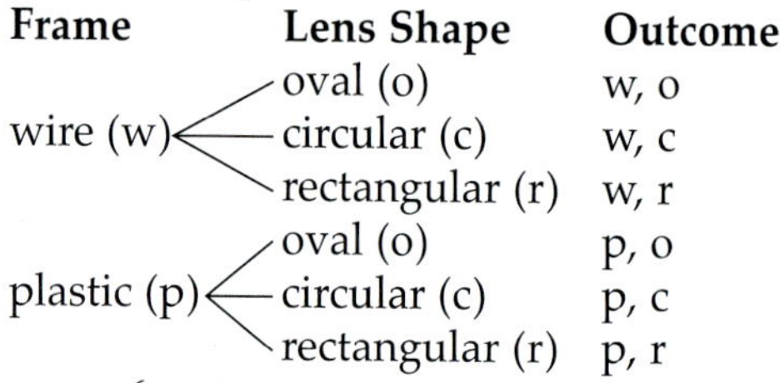

There are 6 outcomes.

Page B43 Review Lesson 6-1

6b.

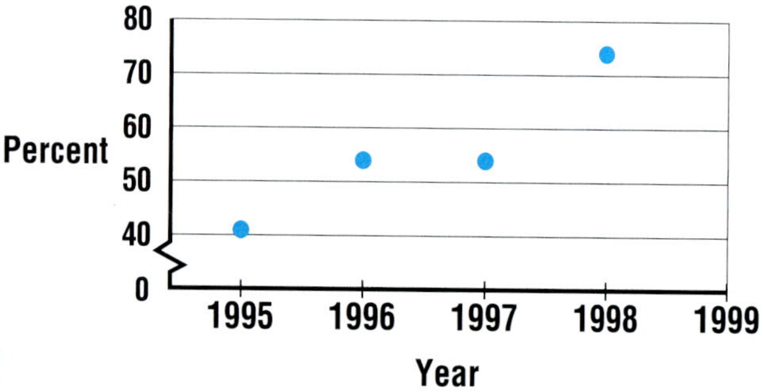

Page B44 Review Lesson 6-2

9b.

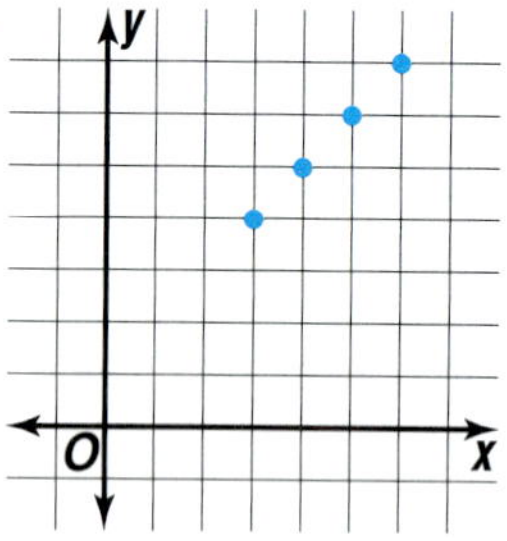

Page B45 Review Lesson 6-3

4.

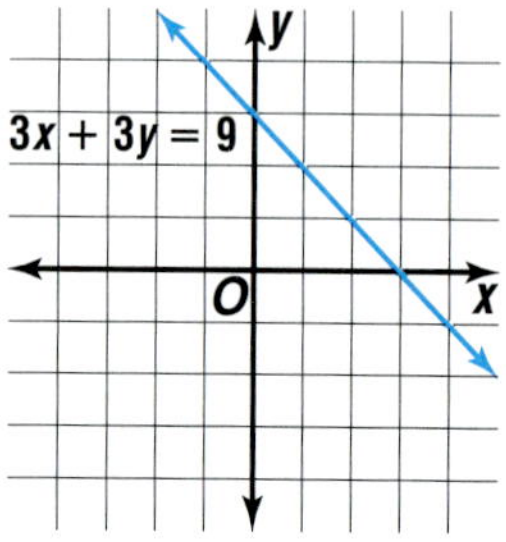

5.

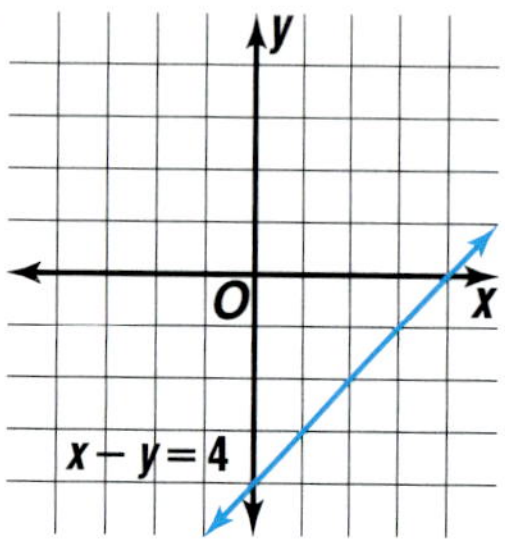

6.

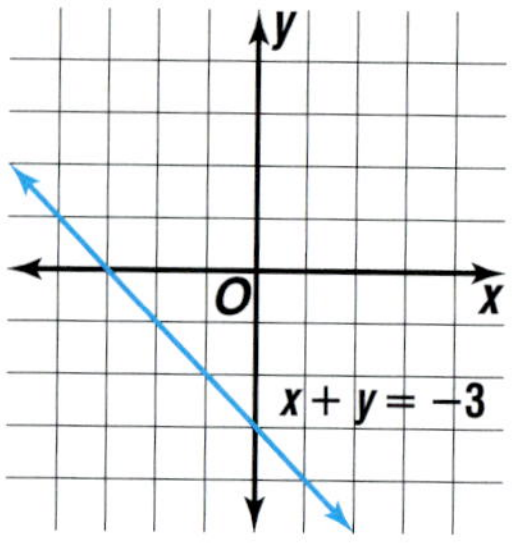

7.

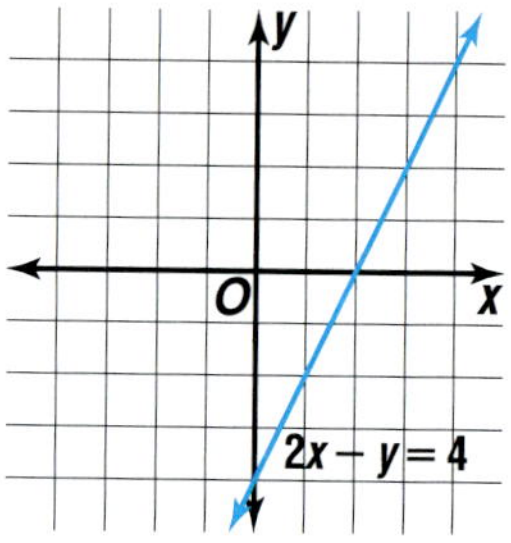

8.

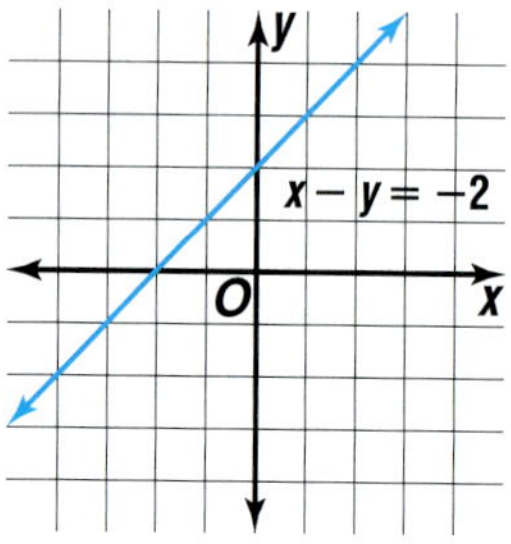

9.

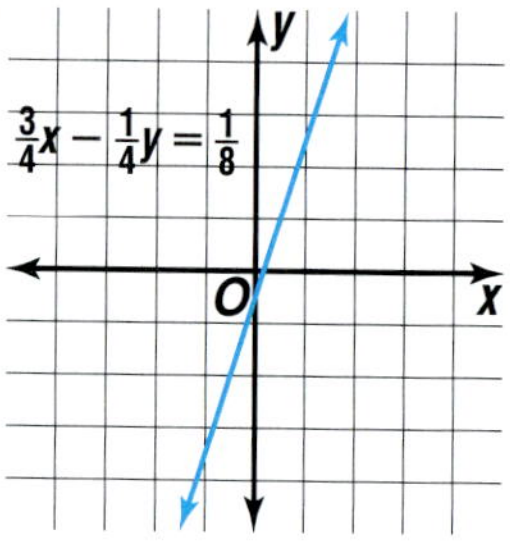

10a.

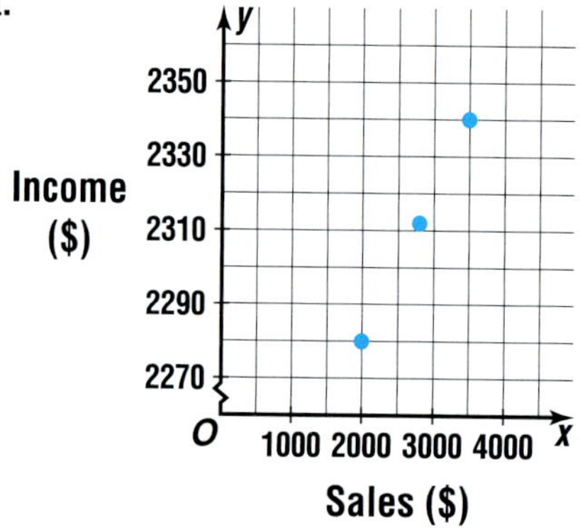

Page B46 Review Lesson 6-4

16a.

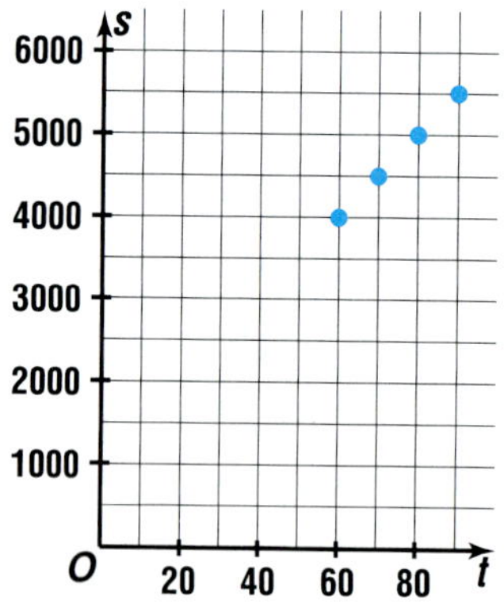

Page B47 Review Lesson 6-5

1. yes

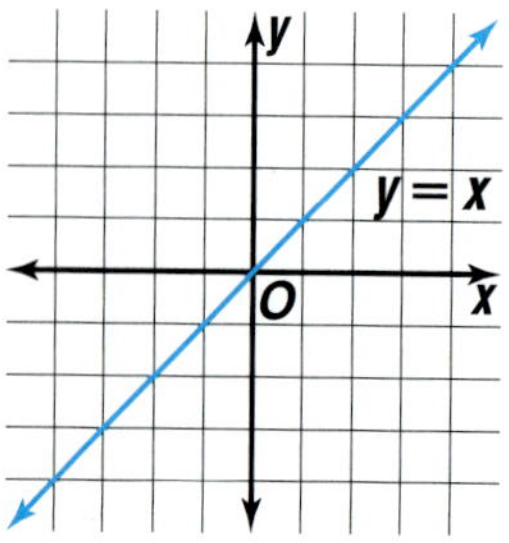

2. yes

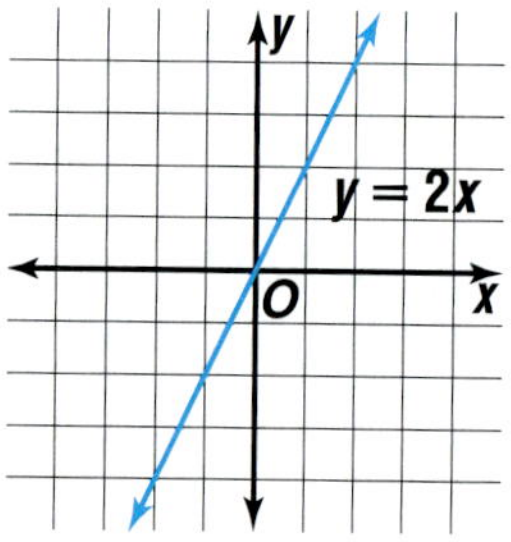

3. no

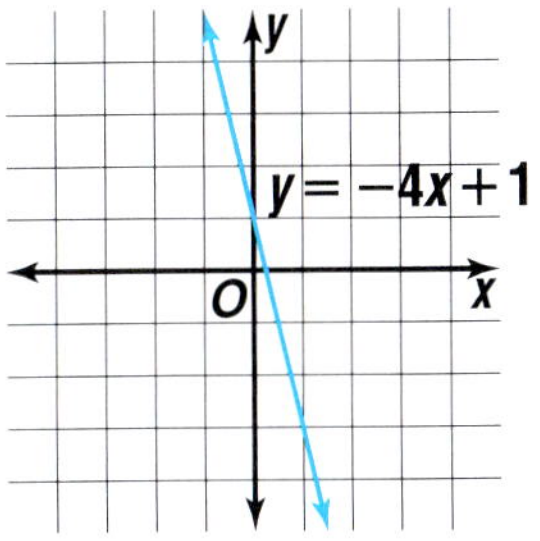

4. yes

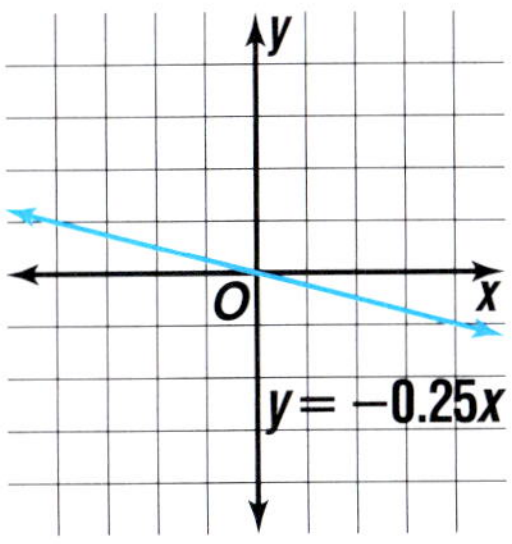

5. no

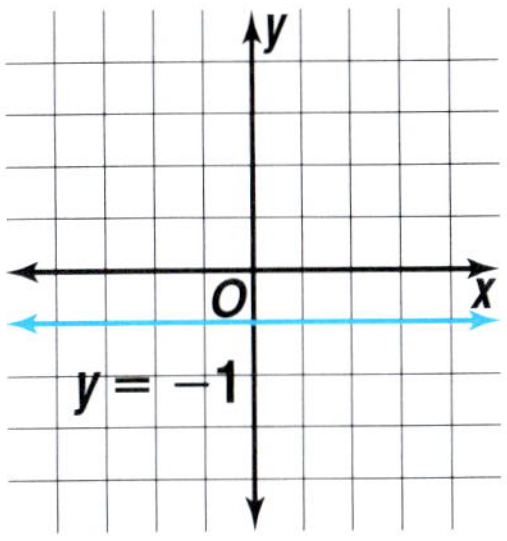

6. yes

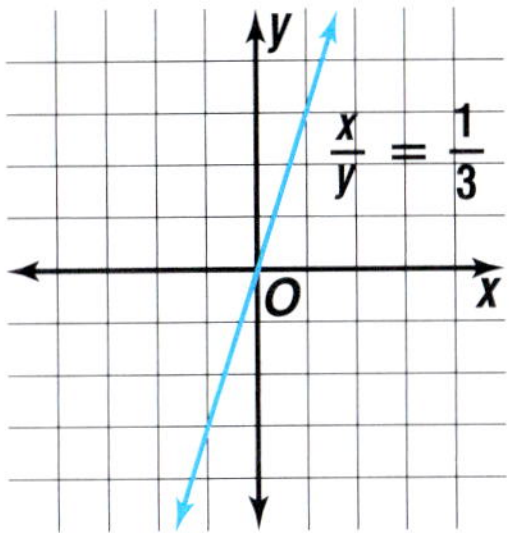

Page B49 Chapter 6 Test

7.

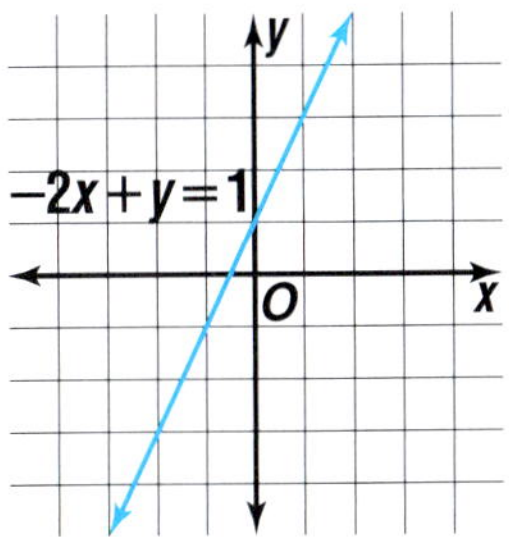

8.

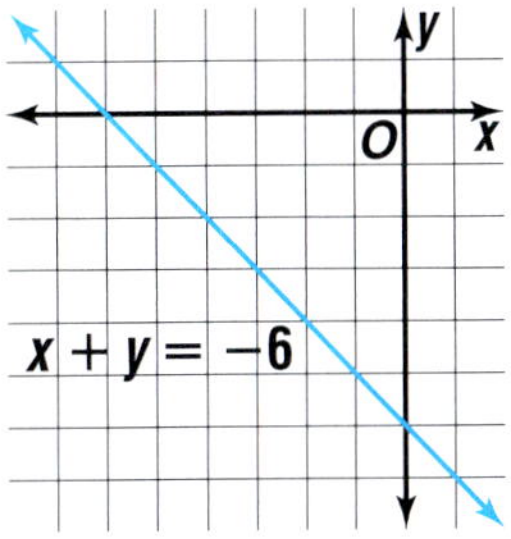

9.

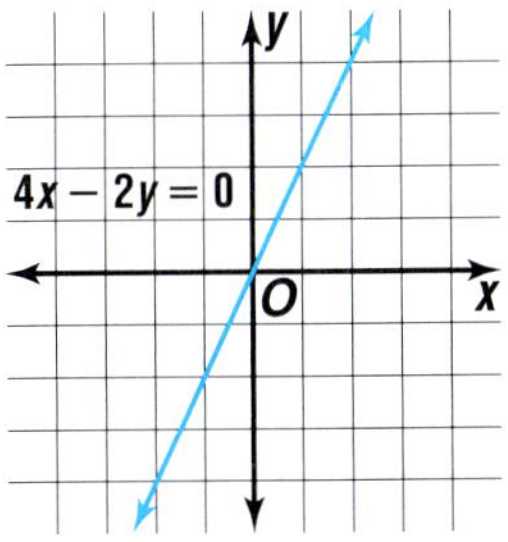

20c.

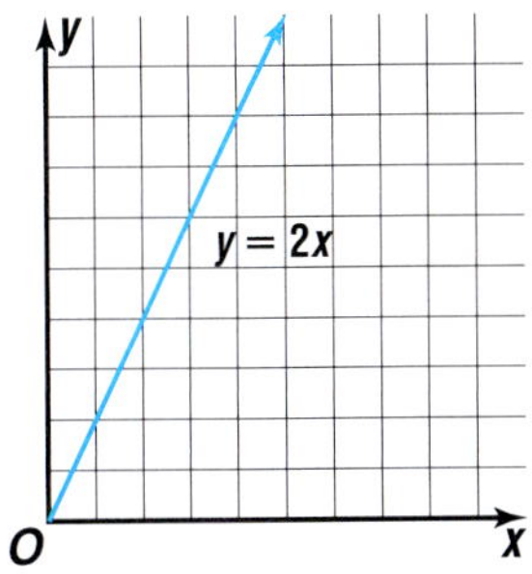

Page B53 Review Lesson 7-4

1a.

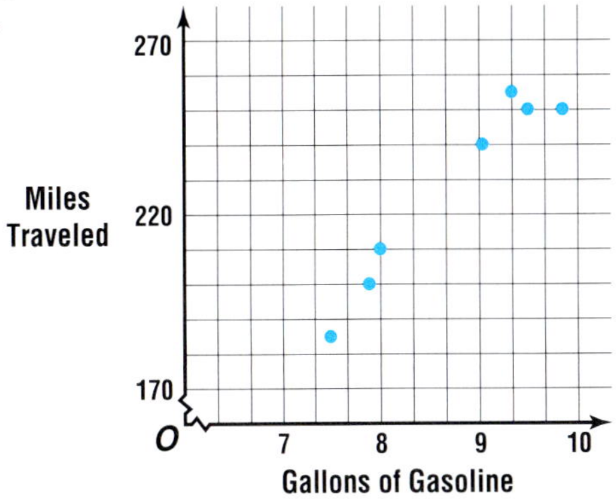

2a.

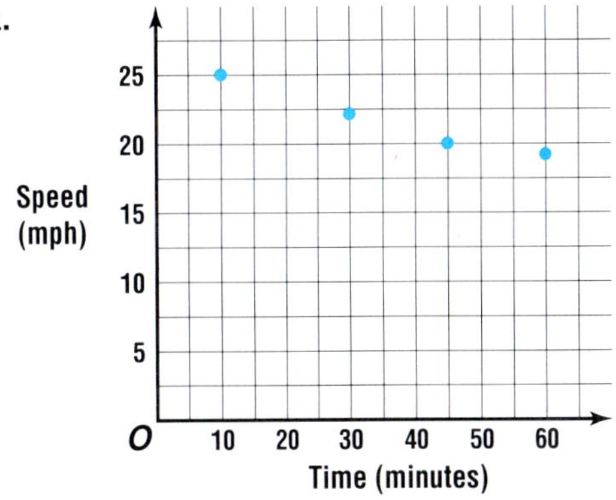

3a.

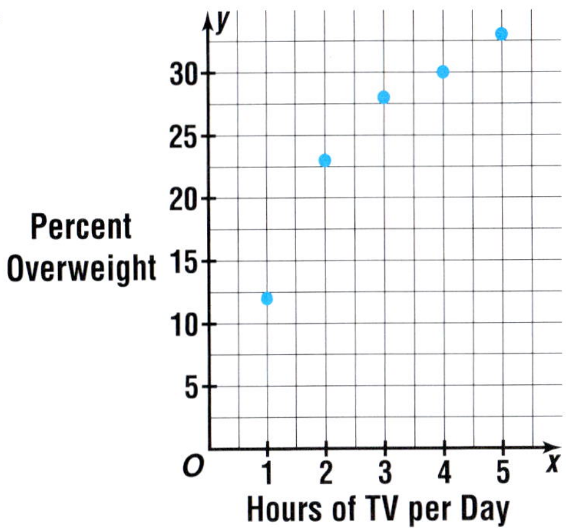

Page B54 Review Lesson 7-5

1. 5, −3

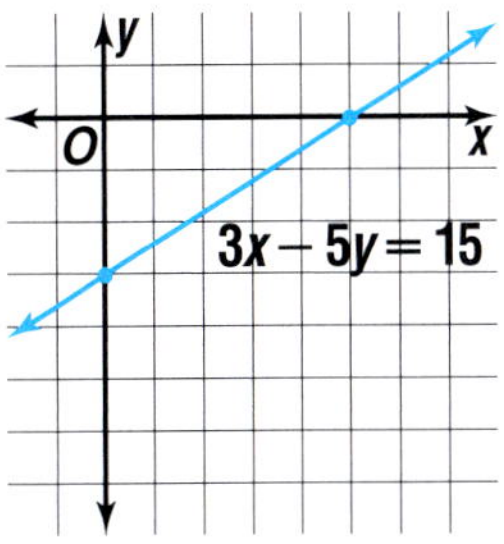

2. 14, 2

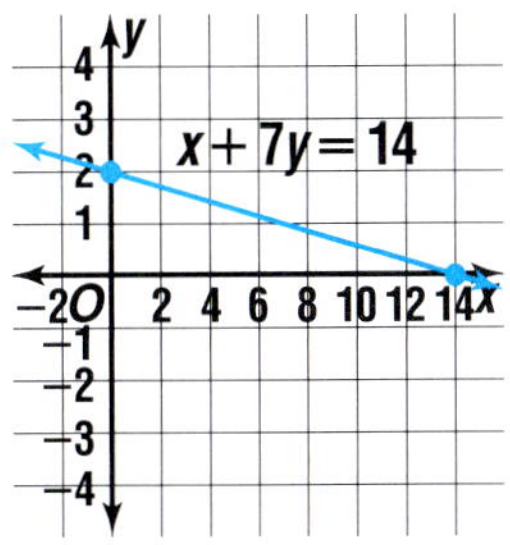

3. $m = \frac{1}{5}, b = 2$

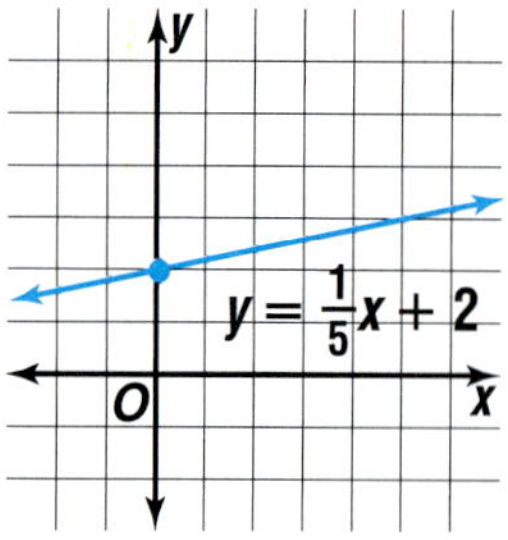

4. $m = \frac{3}{4}, b = -3$

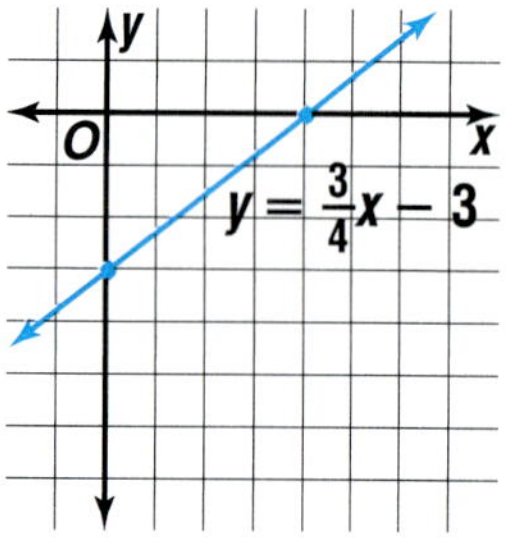

5a.

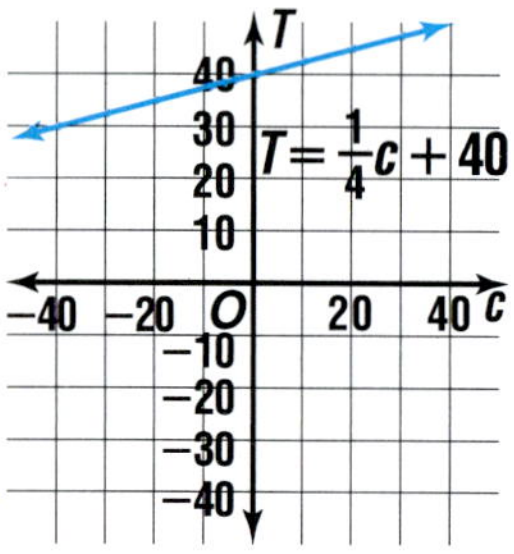

Page B55 Review Lesson 7-6

1.

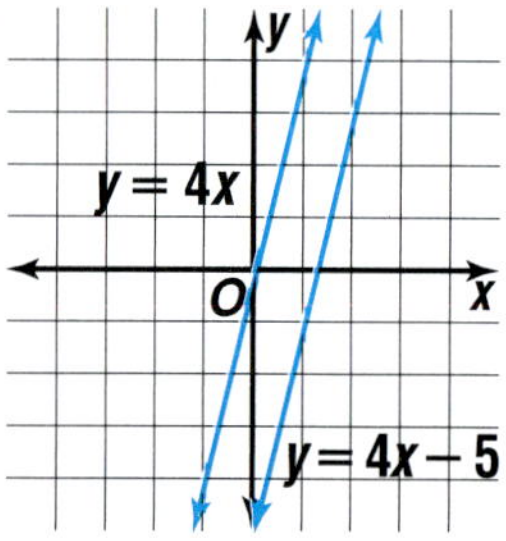

same slope different y-intercepts

2.

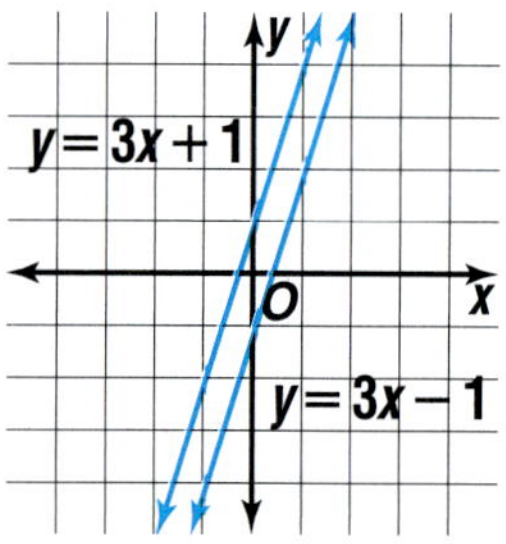

same slope different y-intercepts

3.

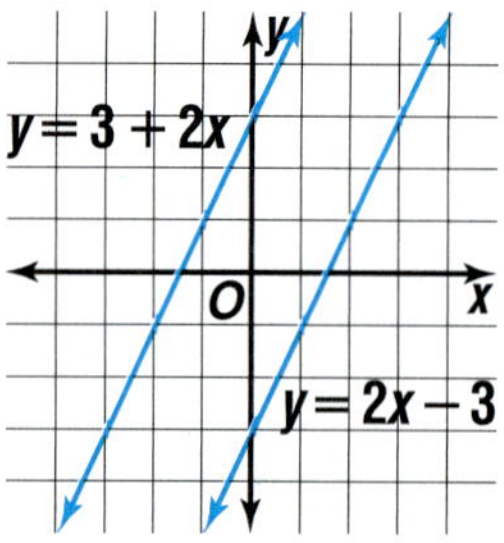

same slope different y-intercepts

4.

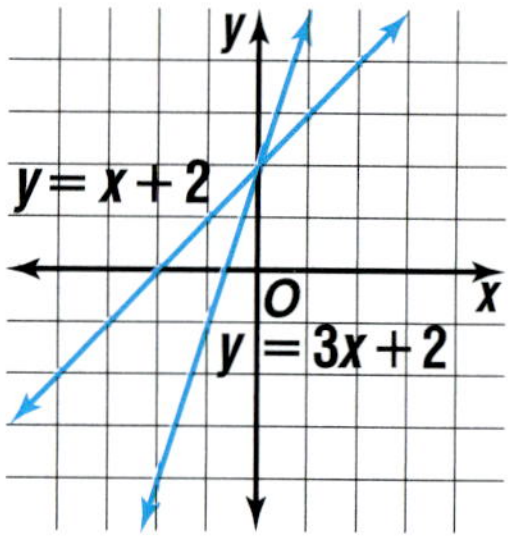

same y-intercept different slopes

5.

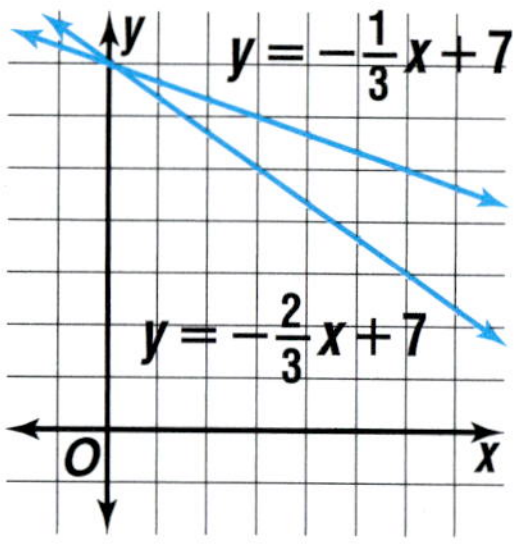

same y-intercept different slopes

6.

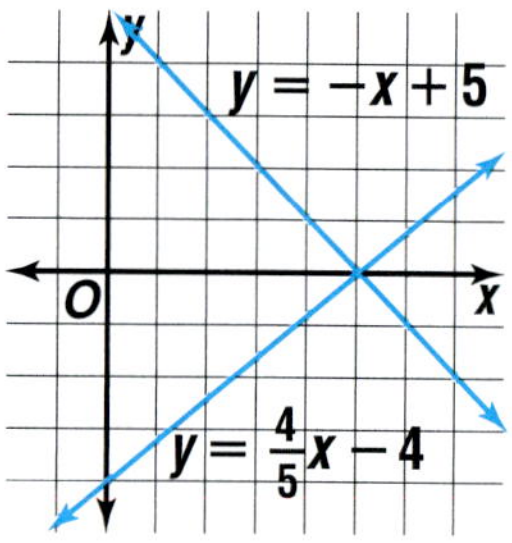

same x-intercept different slopes different y-intercepts

Page B57 Chapter 7 Test

8.

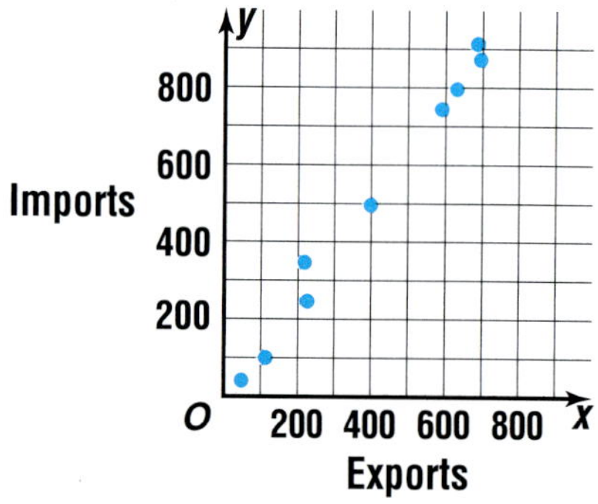

13.

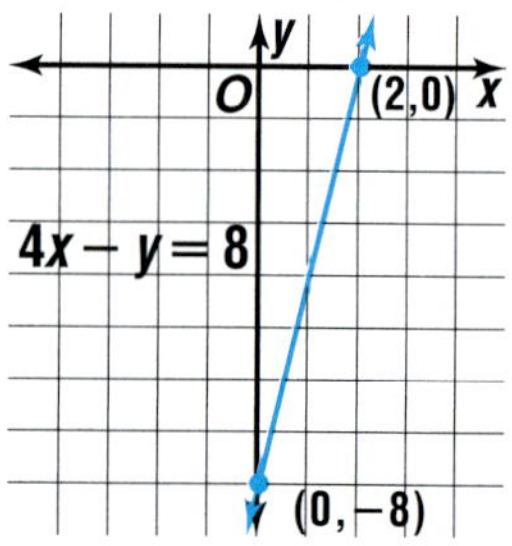

14.

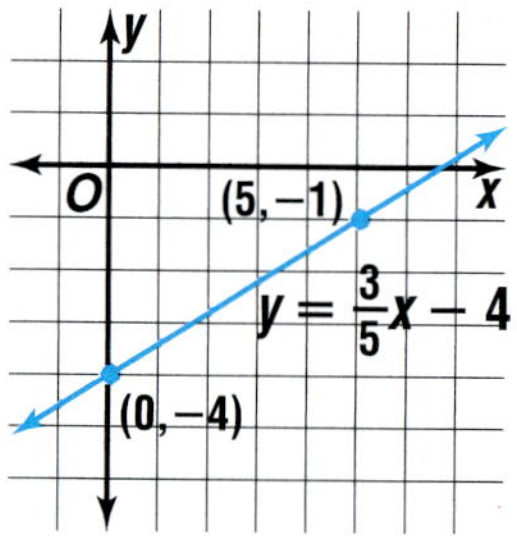